ASE Lab Books

CHROMATOGRAPHY

ASE Lab Books
General Editor A. A. Bishop

Biology editors D. G. Mackean and A. Davies
Chemistry editors Miss F. Eastwood and E. H. Coulson
Physics editors D. Shires and M. F. James
Middle Schools editor Eric Deeson

Already published

BIOLOGY
Plant Physiology compiled by C. J. Clegg
Ecology compiled by A. Davies
Cytology, Genetics and Evolution compiled by G. W. Shaw

CHEMISTRY
Chemical Equilibrium, Acids and Bases compiled by J. M. Newman
Chromatography compiled by R. Worley

MIDDLE SCHOOL
Physics and the Earth Sciences for Middle Schools compiled by E. Deeson
Biology and Chemistry for Middle Schools compiled by J. Bushell

In preparation

CHEMISTRY
Energy and Chemistry compiled by W. J. Hughes

PHYSICS
Mechanics and Properties of Matter compiled by J. C. Siddons
Light compiled by E. Deeson
Heat compiled by W. K. Mace
Electronics compiled by E. W. Mackman

ASE Lab Books

CHROMATOGRAPHY

Compiled by
R. WORLEY

John Murray   Albemarle Street   London

ACKNOWLEDGEMENTS

I would sincerely like to thank
Frances Eastwood and Kit Swinfen
for all their help in producing this booklet.

Printed in Great Britain by Butler & Tanner Ltd,
Frome and London

0 7195 2883 6

Contents

Paper chromatography

Paper chromatography for junior schools

A. V. JONES

The materials used were water-soluble coloured dye mixtures, the main solvent was water, and the types of paper used were student-grade filter paper, newspaper or any other suitable absorbent paper.

WHAT IS CHROMATOGRAPHY?

Paper chromatography is a method of separating and identifying substances by soaking a solvent along a sheet of absorbent paper, which can be supported horizontally or vertically. Because there are many variations of the chromatographic techniques, only a limited number of experiments will be described fully.

HORIZONTAL PAPER CHROMATOGRAPHY EXPERIMENTS

I. A circular student-grade filter paper is placed on a clean, dry surface and a spot of a water-soluble magic marker is placed at the centre of the paper (black or brown 'Scripto' felt-tipped pen works well). A tongue is cut across the paper to within about 1 cm of the spot and the tongue is folded down at right angles.

A container (jam jar, yoghurt cup or egg cup, etc.) is filled to about 2 cm from the top with water (the higher the water level the quicker the solvent soaks the paper and reaches the spot). The paper is placed on top of the container so that the tongue dips into the solvent pool (see Fig. 1).

The components of the mixture separate out as the solvent progressively soaks the paper. To ensure that the solvent does not evaporate from the paper before it has time to soak it, the whole apparatus is enclosed in a container such as a 'Pyrex' mixing bowl, plastic bag, bucket etc. (obviously a transparent container would be the more suitable).

The experiment is allowed to continue until the solvent front reaches about 1 cm from the edge of the paper, or until a suitable time has elapsed.

When the paper is removed it is necessary to dry it as soon as possible, otherwise the separated components would tend to merge back together again. Drying can be done with a hair dryer, infra-red lamp, over a radiator, etc.

The dry chromatogram containing the separated rings of coloured components is usually marked with a pencil around the colours in case they fade.

The chromatograms can be labelled with a pencil, giving the name of the mixture, the solvent, position horizontal or otherwise, and the date.

If the components are weak in colour and difficult to locate then the paper can be placed on a window or television screen as this helps to show up the paler colours.

II. In order to allow for more than one coloured substance to be used on one filter paper at a time, the paper can be cut with slits radiating from the centre. This prevents the colours from merging, provided the spots are placed inside the V of the slits. A wick of cotton wool can be used instead of a paper tongue

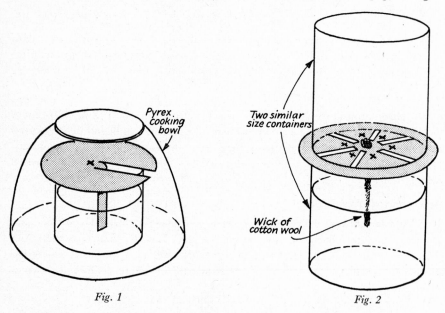

Pyrex cooking bowl

Two similar size containers

Wick of cotton wool

Fig. 1   *Fig. 2*

and this ensures a quicker transfer of solvent to the paper (Fig. 2). Two similar sized containers, one on top of the other and enclosing the paper between them, can also be used to prevent solvent evaporation.

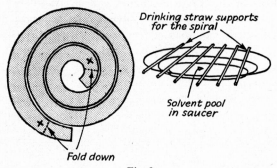

Drinking straw supports for the spiral

Solvent pool in saucer

Fold down

Fig. 3

III. Another variation of the horizontal method makes use of a spiral cut from the circular student-grade filter paper, which is supported on a rack of drinking straws covering a saucer of solvent. The edges of the paper must not touch (Fig. 3). The end, or centre, or both ends of the paper dip into the water in the saucer from a position just behind the spot of magic marker. In this case the solvent pushes the components around the spiral. It is interesting to compare the degree of separation at each end of the paper.

IV. As an alternative to the spiral shape of III, other miscellaneous shapes can be used. This can be the beginning of an open-ended investigation, as the shapes can be to the pupils' design. S, V, U, W, Z, L and D shapes have proved interesting (Fig. 4).

Fig. 4

V. A large square of paper can be spotted many times along a pencilled base line and may be supported on cotton reels, corks or drinking straws. The width of the paper will be determined by the size of the solvent receptacle. This method is very useful for comparing the coloured dyes of sweets of different brands, or of the full range of coloured felt-tipped pens.

The procedure for a suitable experiment using Smarties would be to place the sheet of paper on a clean, dry surface and draw a pencil line across the paper about 6 cm from one end. The Smarties are licked and the sweet lightly dabbed on to the paper to make a small spot. To make a suitably concentrated spot the procedure can be repeated several times with the same sweet. The spots from the separate sweets should be evenly spaced across the pencilled line (Fig. 5).

The distance moved by any component in the dye mixtures will be at the same position from the base line at any particular time. The solvent front is allowed to travel within a few centimetres of the end of the paper and then the papers are dried and the spots circled with pencil. It is interesting for the children to compare the dyes of various sweets, lollipops, etc.

Unless the experiments are repeated identically they cannot be accurately compared. Although the absolute position of the dyes are difficult to reproduce, the position of the components relative to the distance moved by the

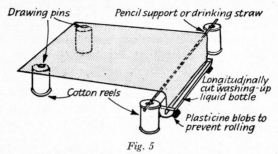

Fig. 5

solvent is found to be a constant value, called the R_f value, and this holds true for short or long distances moved by the component, although it is more accurate to measure these values when the distances are fairly large.

$$R_f = \frac{\text{distance moved by the component}}{\text{distance moved by the solvent}}$$

Provided the solvent moves evenly in this experiment there is no danger of the spots merging together, so no slits need be cut to separate the spots.

In all the experiments I to V it is desirable (but not absolutely necessary if using water for a short length of time) to enclose the whole apparatus in a suitable transparent, air-tight container.

The process of drying, locating and labelling is common to all methods used.

VERTICAL CHROMATOGRAPHIC METHODS

There are many applications of chromatographic separations and often the vertical methods can achieve separations not possible with horizontal methods. The principles of spotting, enclosure, drying, location and labelling are the same as with the horizontal methods.

The vertical methods can be subdivided into ascending methods when the solvent pool is placed at the base of the paper, and descending methods when the solvent is allowed to soak from the top of the paper downwards.

ASCENDING METHODS

The paper is spotted about 3 cm from the bottom of the sheet and this is submerged in the solvent to a depth such that the spot is about 1 cm above the solvent surface. (Under no circumstances must the spot be below or in direct contact with the solvent pool as the spot would simply dissolve off the paper into the solvent.)

A(1). Place a spot of dye mixture on a strip of absorbent paper about 2 cm from one end. The paper can be filter paper or a clean white strip cut from the edge of a newspaper. Place about 1 cm of water in the bottom of a jam jar or milk bottle and cover with a lid. Push the strip of paper through a slit in the lid until it dips about 1 cm into the water, making sure that the paper does not touch the sides of the container (Fig. 6).

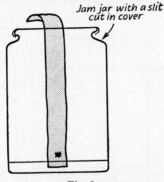

Jam jar with a slit cut in cover

Fig. 6

The solvent moves up the paper quickly at first, pushing the components of the mixture with it; but the solvent moves progressively more slowly.

Usually an experiment of this type would take from 30 minutes to an hour to reach within a few centimetres of the top of the paper.

The main difficulty of the method is that solvent flow can only occur until it reaches the top of the paper, when the flow ceases and so separation stops and substances which move slowly (low R_f value) are often incompletely separated.

Longer lengths of paper can be used if taller containers are employed, e.g. kilner jars, gas jars or buckets.

A(2). A simple, cheap but effective piece of apparatus suitable for ascending chromatography is almost completely made from a plastic washing-up bottle (Fig. 7).

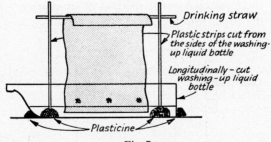

Drinking straw

Plastic strips cut from the sides of the washing-up liquid bottle

Longitudinally – cut washing-up liquid bottle

Plasticine

Fig. 7

The longitudinally cut bottle is prevented from rolling by supporting with a wad of plasticine. Plasticine is also used to support the side arms which are cut from the same plastic bottle. In the groove of the side arms is placed a pencil, or tube, or two interlocking drinking straws, to form a bar across the top. To this bar is attached the square piece of filter paper, which has been spotted along the base. The paper should just dip into the solvent in the washing-up bottle container. The whole apparatus is easily enclosed in a plastic bag. The experiment is stopped when the solvent front reaches to within 1 cm of the top of the paper.

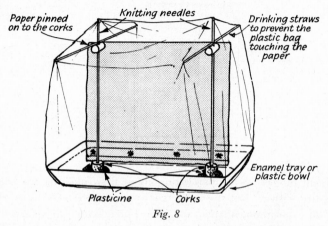

Paper pinned on to the corks   Knitting needles   Drinking straws to prevent the plastic bag touching the paper

Enamel tray or plastic bowl

Plasticine   Corks

Fig. 8

A(3). The apparatus shown in Fig. 8 consists of a supporting frame of corks pierced by metal or plastic knitting needles, and kept in position in the solvent tank by corks embedded in plasticine. The square of chromatography paper is pinned to the top corks so that it dips into the solvent to a depth of about 1 cm. The solvent container can be an aquarium tank, or plastic bowl. A plastic bag can be used to enclose the whole apparatus, but to ensure that the bag does not touch the paper and interfere with the solvent flow, drinking straws are centrally pierced and mounted horizontally at the top of the needles. If necessary, two pieces of paper can be used, pinned to both sides of the corks.

A(4). Possibly the most convenient method of supporting a large square of paper is to roll the paper into a cylinder, after spotting along the pencil line

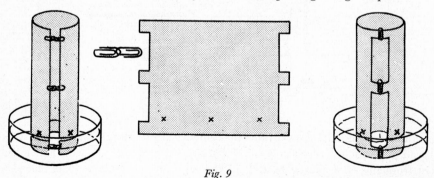

Fig. 9

about 3 cm from the bottom. The edges of the paper are clipped together with pairs of interlocking paper clips, or alternatively, the paper can be cut to leave three tongues on each side opposite each other and these meet when the cylinder is rolled and the tongues are clipped together with paper clips (Fig. 9).

The cylinder will stand up by itself in the bowl or dish containing the solvent. The top of the cylinder of paper can be pierced with two drinking straws and these keep the plastic bag, used to enclose the apparatus, away from the paper.

The experiment is stopped after a suitable time and the cylinder removed from the solvent. The paper is opened out and dried as quickly as possible, the spots of the separate components marked, and, if necessary, the R_f values calculated.

DESCENDING CHROMATOGRAPHY METHODS

In these methods the solvent pool must be supported on top of the paper and this is often difficult to achieve. The basic principles of spotting, drying, enclosing, etc., are also applicable to these methods. A disadvantage of this process is that it is not self stopping, and unless it is watched carefully the separated spots will drip from the end of the paper. This disadvantage can be turned into an advantage when dealing with materials that need a long length of paper to fully separate the components of the mixture, or for components which do not separate very quickly. In this method the process will continue at a steady rate and not slow down like the ascending methods. The strip of paper above the spot must be made longer than in ascending methods to allow the paper to dip into the pool. In the bottom of the container is placed about 1 cm of solvent (water), which is needed to maintain a saturated water vapour in the container.

D(1). The simple piece of apparatus shown in Fig. 10 easily demonstrates the descending method. The solvent pool is the wad of soaked cotton wool covering the end of the paper and is placed just behind the slit in the cover (if placed on top of the slit there might be a tendency for the solvent to flow down the paper instead of soaking down).

The cotton wool pad can be resoaked when the solvent has been used up. The pad is covered with an inverted yoghurt cup to minimize solvent evaporation. A small quantity of solvent in the bottom of the jam jar ensures a saturated solvent vapour around the paper and prevents evaporation from the paper.

D(2). The top cut from a washing-up liquid bottle, when inverted and slit above the solvent pool level, makes a suitable descending chromatography pool. The receptacle can be a jam jar or the remaining part of the washing-up liquid bottle. The glass container is obviously more useful as the experiment can be more easily followed (Fig. 11).

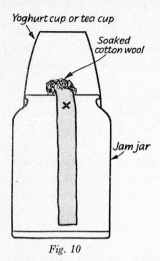

Yoghurt cup or tea cup

Soaked cotton wool

Jam jar

Fig. 10

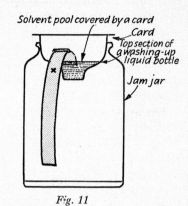

Solvent pool covered by a card

Card

Top section of a washing-up liquid bottle

Jam jar

Fig. 11

D(3). A longer piece of paper can be used when a kilner jar is employed. An even longer chromatogram can be run by using a bucket or a glass case made from glass sheets hinged with 'Elastoplast' or 'Sellotape' (Fig. 12).

D(4). In this descending method a wad of soaked, but not too wet, cotton wool is folded over the top of a strip of paper and the paper clipped inside a plastic bag. The bag is inflated to prevent the sides touching the paper strip and it acts as a suitable enclosure to maintain the saturated vapour around the paper (Fig. 13).

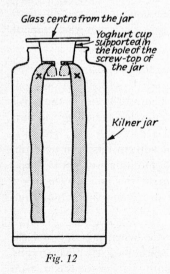

Glass centre from the jar

Yoghurt cup supported in the hole of the screw-top of the jar

Kilner jar

Fig. 12

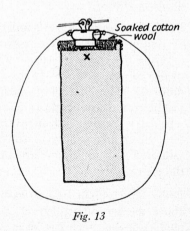

Soaked cotton wool

Fig. 13

An extremely long strip of paper can be used if a large dry-cleaner's bag is used. The clip can be hung from a shelf or hook.

TWO-WAY CHROMATOGRAPHY

In the experiments using a square of paper, when surveying the dried chromatogram it was noticed that frequently there were incomplete separations or overlapping coloured spots. In an endeavour to achieve a complete separation the following adaptation can be made.

The square of paper is pencil-ruled in two directions at right angles about 5 cm from each edge, and the paper spotted at the intersection of these lines (only one dye mixture can be used at a time). A chromatographic separation is obtained in one direction as previously described. The paper is dried and turned through 90° and then placed in a tank of a different solvent and the chromatogram again allowed to develop and stopped at a suitable time, dried and labelled. The spots which were previously incompletely separated will now become completely separated as it is very improbable that two substances will have identical characteristics and properties in two different solvents. Water can be used for one direction and methylated spirits, acetone, dilute acids, etc., for the other direction.

MISCELLANEOUS METHODS

There are some experiments which can be devised that cannot be easily classified under the preceding headings, but they are still of interest. Consider, for example, the experiment which makes use of a roll of stiff metal foil (Fig. 14).

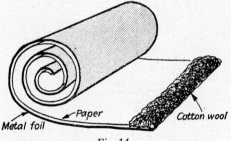

Fig. 14

A roll of zinc, copper or stiff aluminium is loosely coiled and a piece of paper fed slowly into the roll until it has reached the end of the coil. The protruding end of the paper can be wetted with a wad of cotton wool soaked in solvent, or the end can be folded into a suitable solvent pool. The method makes use of ascending, descending and horizontal methods, all in one piece of apparatus. The contact between the moist paper and the metal surface does not adversely affect the chromatograms, and the solvent does not readily evaporate from the paper. At the end of the experiment the paper can be removed by unrolling the coil, or alternatively, the paper can be left in the roll and dried with a hair-dryer in the coil.

Another method for carrying out descending paper chromatography follows:

Test-tube chromatography

L. O'DONNELL

An inexpensive method which enables individual work on descending paper chromatography to be done in $15 \times 2\frac{1}{2}$ cm boiling tubes is shown.

The strip of chromatography paper is folded over the small glass tube and the upper and lower paper-clips placed in position. The paper is spotted just below the upper clip. The ends of the short glass tube are placed in the supporting staples fixed in the cork. The 'tail' of the paper is folded up to the cork which is then put in the tube, the loop of paper being adjusted to a convenient size. The glass tube containing cotton wool packed for two-thirds of its length should be a loose enough fit in the cork to allow it to be

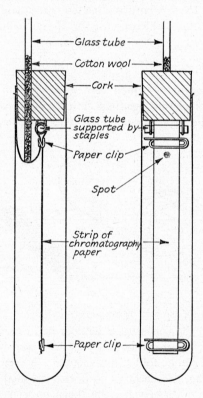

adjusted so that the cotton wool in the lower end just touches the loop of paper. The solvent is poured through a narrow funnel into the glass tube. The volume of liquid needed varies with the solvent.

If the paper-clips are almost as wide as the tube they will prevent the paper from touching the sides of the tube. The type of paper-clip which is shaped like the letter α, when opened out, makes a very convenient staple.

Paper chromatography of felt-tip pens

J. W. CLARIDGE AND B. L. CLARIDGE

The separation of the coloured dyes in writing ink such as 'Quink' black by paper chromatography has proved to be a popular experiment with younger forms, especially when the scope of the experiment can be widened. This we have done by discovering that the multitude of coloured felt-tip pens which children carry with them often produce interesting chromatograms. The 'Gem' watercolour type is one of the most suitable, and can be eluted with water.

In the interests of speed, simplicity and economy, we have rationalized the experiment to the point where even the least coordinated child can produce acceptable results quickly and cheaply.

The apparatus consists of a plastic yoghurt beaker containing 1 cm depth of water, and half a wooden splint. The pupil is provided with 10 cm $\times$ 1 cm strips cut from a sheet of white chromatography paper, and is instructed to make a dot with a felt-tip about 1 cm from one end. The other end must then be wrapped tightly around the splint sufficiently to allow the strip to hang straight with its end just dipping into the water, when the splint is laid across the beaker. Excellent separation is achieved in a few minutes.

Methods described elsewhere [1, 2] use filter paper circles or sticks of chalk, both of which become costly items when 120 children each want to make half a dozen or more chromatograms. Our method is cheap, and gives superior results in considerably less time.

REFERENCES

1. Nuffield Chemistry Sample Scheme, pp. 22–5.
2. Vernon, E. J., *Chemistry, an Introduction* (Chatto & Windus, 1967), p. 56.

Simple experiments in chromatography

W. J. OWEN AND G. AGAR

Some experiments in chromatography have been worked up with the junior forms in mind. These have been considered with respect to simplicity.

Three solvents have been used: water, ethanol and propanone (acetone) and the results obtained with various dyes and indicators have been most encouraging.

TECHNIQUE

One spot of indicator is put in the centre of a Whatman No. 1 circular filter paper and is allowed to dry, keeping the paper flat. One drop of water is added using a glass rod, and is left for 30 sec (observe). Repeat adding water slowly, dropwise. Separation into concentric rings is easily seen. Screened methyl orange shows such results very well.

A more sophisticated technique is to place filter paper horizontally between two flat glass plates, the upper plate having a small aperture in the centre, with the centre of the paper opposite the hole. One drop of dyestuff is introduced and is allowed to dry. Steady dropwise addition of water again shows separation into concentric rings.

Perspex plates could be used instead of glass, but they may need clamping.

This has been developed further by using upward eluting technique with various inks.

A simple class experiment is as follows: three separate boiling tubes are used, each set up as shown in Fig. 1.

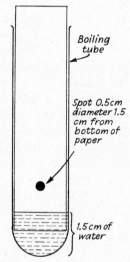

Fig. 1. *Experimental set-up: three tubes arranged as shown*

A—'Quink' black ink.
B—Red ink.
C—A mixture of red and black ink.

Care must be taken to ensure that the sides of the tube are dry, hence water is introduced by wash bottle or pipette. On the introduction of the Whatman No. 1 chromatography paper, it is important that the edge is evenly introduced to the water level.

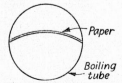

Fig. 2. *Top view of boiling tube arrangement*

N.B. The paper must be slightly broader than the internal diameter of the boiling tube.

A and *B* can be used as known samples, and *C* can be used as an unknown. The pupils are asked to comment on their observations.

Separations can be seen after as little as 3 minutes and after 25 minutes accurate observations and deductions can be made.

If the spot is too large then capillary action between the side of the tube and the paper will lead to false results.

Water is found to give the best separations, and is easily workable in a practical period.

Further work has been tried using other dyestuffs and indicators with different solvents (Fig. 3).

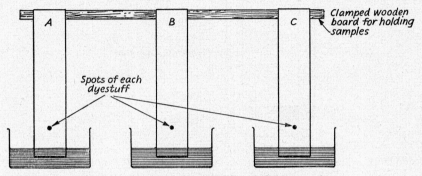

Fig. 3. Experiment for comparison and demonstration

The above demonstration may also be useful for comparisons during the practical work of the pupils.

Separation of laboratory indicator dyes

D. J. JEWELL

Solutions of congo-red, bromophenol-blue and phenol-red, containing 80 mg of dye in 100 cm³ of ethanol, are prepared together with a mixture of all three at the same concentration. The apparatus shown in the diagram is set up, with a pencil line drawn one inch from the bottom of the paper. Four equidistant points are marked on this line using a pencil. A pencil must always be used to mark chromatography paper as inks will run in the solvents. The indicators are applied using a platinum wire bent into a small loop at its end. The wire should be cleaned between each application by heating to redness in a Bunsen flame. Three or four drops of each indicator are applied and it is most important to keep the diameter of the spot under ½ cm. This can be done by drying the spot between each application with a 'Morphy-Richards' type of hair-dryer. The solvent-system for this separation is a mixture of butan-1-ol/ethanol/2 M ammonia (60 : 20 : 20); it is poured into the bottom of the jar to a depth of about 2 cm. The paper is then fumed with 0.880 ammonia to convert the indicators to their basic form and is then allowed to hang above the solvent for about 15 minutes, after the jar has been

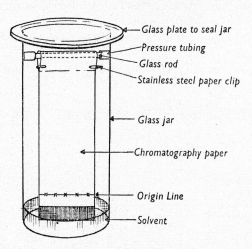

sealed with a glass plate. The glass rod is then lowered until the paper dips $\frac{1}{2}$ cm into the solvent. The applied spots must not be allowed to dip into the solvent, otherwise the dyes will wash off. The jar must be kept sealed throughout the separation to prevent evaporation losses from the solvent. Separation is rapid, starting as soon as the solvent front reaches the coloured spots, and is virtually complete after 90 minutes. The experiment may be repeated using a second solvent-system of water/saturated ammonium sulphate(vi)/ethanol (75 : 10 : 15).

REFERENCE

Smith, I., *Chromatographic Techniques* (Heinemann).

Some uses of chromatography in botany

W. M. M. BARON

1. PAPER CHROMATOGRAMS USING PHENOL AS SOLVENT

(*a*) *Making up the phenol solvent*

Take 28 g of pure crystalline phenol (this should be white and unoxidized). Place this in a stoppered separating funnel and add 12 cm³ of distilled water. Add a little sodium chloride and shake thoroughly. Fill the funnel with coal gas or natural gas to prevent oxidation and leave overnight. The lower layer is the phenol saturated with water. The upper layer is rejected.

(b) The chromatogram jar

A tall jar 45 cm high is fitted with a cork drilled for a glass rod which has a small hook in its end. This glass rod should be able to slide up and down inside the cork. 1 cm of solvent is placed in the bottom of the jar. A long strip of Whatman No. 1 filter paper is then cut for size. It should be able to hang freely in the jar with the glass rod raised so as *not* to touch the solvent.

(c) Loading the paper

Mark a pencil spot $2\frac{1}{2}$ cm from the base of the paper strip and load the unknown or control, known substance, carefully on to the spot using a capillary tube. It is important to load plenty of material (unless it is a concentrated solution of a single known substance) but it is also important to keep the spot less than $\frac{1}{2}$ cm diameter.

(d) Running the chromatogram

Attach the loaded filter paper to the hook, using a paper-clip. Place it in the jar and leave it for ten minutes for the atmosphere inside the jar to equilibrate. Then lower the glass rod so that the paper dips $\frac{1}{2}$ cm into the solvent. Make sure the cork is a proper seal. The solvent will then rise slowly up the paper carrying materials from the spot with it. Different substances are differently adsorbed and are therefore carried different distances up the paper. After about 20 hours (with phenol solvent and a 36 cm paper) the paper is removed and dried carefully at an electric fire, preferably in a fume cupboard.

(e) Development of the chromatogram

The dried paper is then ready for spraying with an appropriate spray to show up the various substances. The spray depends on the substances being investigated:

1. *Plant acids:* Bromothymol-blue pH 8.5 (NaOH added).
2. *Sugars:* 10 per cent benzene-1,3-diol + HCl in propanone (acetone).
3. *Amino acids:* 2 per cent Ninhydrin in butan-1-ol.

For the colours to appear after spraying for sugars and amino acids, it is necessary to heat the paper quite strongly (note that these sprays are very inflammable). Mark the spots in pencil as soon as they appear and also the height the solvent reached.

(f) Calculation of results

It is necessary to calculate the R_f of the spot:

$$R_f = \frac{\text{Height spot from start}}{\text{Total height solvent from start}}$$

The R_f value of a particular substance should always be the same provided the chromatogram is treated in the same manner in each case. Different paper, solvent and running conditions may affect the R_f values. R_f values of commoner substances may be found in tables but it is usually best to run a parallel chromatogram using one of the substances in the pure state that is thought to be present in the mixture. Comparison of R_f's is then much more accurate.

(g) *Tables of R_f values using phenol as solvent*

   (i) *Plant acids*   *Colour of spot*

2,3-Dihydroxybutane-1,4-dioic acid (tartaric acid)	0.28	yellow
2-Hydroxypropane-1,2,3-tricarboxylic acid (citric acid)	0.32	,,
Ethanedioic acid (oxalic acid)	0.42	,,
2-Oxopropanoic acid (pyruvic acid)	0.59	,,
2-Hydroxybutanedioic acid (malic acid)	0.42	,,

   (ii) Sugars

Glucose	0.19	red
Fructose	0.40	,,
Sucrose	0.27	,,

   (iii) *Amino acids*

Glutamic acid	0.38	orange red
Glycine	0.50	brown purple
Arginine	0.70	red purple
Alanine	0.72	blue purple
Tyrosine	0.66	deep purple
Leucine	0.91	,, ,,
Proline	0.95	yellow

(h) *Applications of phenol chromatograms*

   (1) *Plant acids:*   Succulent acid metabolism.
   (2) *Sugars:*   (a) Photosynthetic products
                      (b) Release of simple carbohydrates during germination
   (3) *Amino acids:* Presence in active growing areas compared with dormant tissues.

N.B.   Notes on the extraction of plant acids may be found on page 51.

2. CHROMATOGRAMS USING PETROL-ETHER/PROPANONE (ACETONE) AS SOLVENT

This is particularly suited to investigation of the chlorophyll pigments. It is first necessary to prepare a pure extract of the pigments.

(a) *Preparation of the chlorophyll extract*

Add 50 cm³ of 90 per cent propanone to an equivalent quantity of dried nettle leaves in a mortar and grind to extract the pigments. The extract is then filtered through glass wool at a Büchner funnel and the filtrate placed in a separating funnel, diluted with its own volume of petrol-ether and shaken. An equal volume of distilled water is then added to wash the extract and the watery layer is rejected. The washing is repeated three times. Sodium sulphate(VI) is then added to help break the emulsion down and the extract is left to stand, together with the sodium sulphate(VI) for one hour. The pure extract can then be used for either the filter paper chromatogram or the column chromatogram.

(b) Filter paper chromatogram of the chlorophylls

This is used for the identification of the chlorophyll pigments. The loading and running of the chromatogram is as described for the phenol chromatogram but the solvent is 100 parts petrol-ether to 12 parts propanone. The chromatogram will only take about half as long (about 10 hours) to run. There is no need of a developing spray; R_f values are given below.

(c) R_f values of the chlorophyll pigments

		Colour of spot
Carotene	0.95	yellow
Phaeophytin	0.83	,,
Xanthophyll	0.71	yellow-brown
Chlorophyll a	0.65	blue-green
Chlorophyll b	0.45	green

(d) Column chromatography

This is used for obtaining separate liquid extracts of each pigment so that their individual absorption spectra and fluorescence properties can be examined.

A 1 cm bore glass tube about 36 cm long is plugged with glass wool and tightly packed with aluminium oxide. The petrol-ether/propanone solvent (100 parts : 12 parts) is then drawn through the column using a filter pump. When the column is uniformly saturated, 5 cm³ of the chlorophyll extract is added at the top and allowed to sink down. Fresh petrol-ether/propanone solvent is then added continuously (use a dropping funnel) at the top. This carries the pigments down separating them as it goes. The fractions can then be collected as they appear at the bottom of the column. They can then be examined individually.

(e) Other uses of these techniques

These techniques can also be adapted for investigating pigment changes in leaves during the autumn and the flower pigments.

Enzyme action illustrated by paper chromatography

D. MORGAN

This technique demonstrates the action of β-amylase on starch and invertase (saccharase) on cane sugar by allowing the enzyme and the substrate, in each case, to react on Whatman No. 2 filter paper. The products of the reaction can be located by appropriate indicators. It is presumed that β-amylase hydrolyses starch to maltose (as well as a number of ill-defined substances) at a pH optimum of 6.6; similarly, invertase hydrolyses sucrose to equal amounts of D-glucose and D-fructose at a pH optimum of 4.5.

(a) The chromatographic paper is cut to suitable dimensions, forming a cylinder to fit within the developing jar. The paper is marked with a pencil line 2 cm from the base; seven small circles are also marked along this line at intervals.

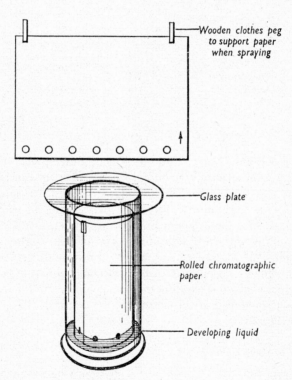

(b) The enzymes, β-amylase and invertase, are made up in 1 per cent aqueous solutions. The starch, sucrose, maltose, glucose and fructose are made up in 0.5 per cent solutions in 25 cm³ amounts of distilled water. A mixture of small amounts of β-amylase and starch, invertase and sucrose is prepared.

(c) One drop of each of the above substrates, end-products and enzyme–substrate mixtures is placed on the marked spots allocated on the filter paper. Each spot is allowed to dry and the process repeated once or twice.

(d) The paper is rolled and held in position as a cylinder with paper clips (preferably of plastic).

(e) The paper is arranged in the jar with the spots lowermost so that development by ascending chromatography will result. The developing solvent-mixture is butanoic acid, butan-1-ol and water in proportions 2 : 1 : 1 at a depth not exceeding 1 cm. A tight-fitting glass lid is required to seal off the jar. The apparatus may be left, thus assembled, for 3–6 hours.

(f) The paper is removed from the jar, opened out flat and allowed to dry in warm air, preferably in a fume cupboard.

(g) The sugar spots and the results of hydrolysis can be detected by spraying the paper lightly with the following mixture:

5 g NH_4Cl dissolved in a solution of 20 cm^3 10 per cent $(NH_4)_2MoO_4$ in 3 cm^3 concentrated HCl aq. Heat the sprayed paper to 70–80 °C by a hot radiator or on a hot-air oven. Carbohydrate areas develop a molybdenum blue colour.

Other indicators may be found on pages 20–21 and in *SSR*, 1957, **136,** 452.

(h) The ensuing end-products of the enzymic hydrolyses can be determined from the R_f values of the known carbohydrates used on the paper.

REFERENCE

Aronoff and Vernon, *Arch. Biochem. Biophys.*, 1950, **28,** 424.

Studies relating to *Capsicum annuum* (red and green peppers)

S. J. HODGE AND B. R. WEARE

Capsicum annuum is a plant which can be easily grown; seeds may be sown from February onwards, and the rather straggling plant thrives in a warm atmosphere with plenty of water. White flowers appear in July or August, and these produce berries, which are green to begin with but ripen to red in September. The plant grows to about 20 cm in height, and produces berries with a length of at least 7 cm. The ripening process may be shown to be photo-dependent, by covering a berry with an opaque material that hinders the colour change.

The pigments were extracted from the berries by crushing them in a liquidizer and extracting the resultant pulp with trichloromethane. The trichloromethane extracts were dried with magnesium sulphate(VI) (anhydrous), and the trichloromethane removed by evaporation on a water bath.

Red and green oils were obtained which crystallized slowly in a refrigerator. The red oil was stable in light, but the green oil was very photo-sensitive. Chromatograms were obtained of the berry pigments, and it was found that the green pigment changed to red while being run on the paper in the presence of light, producing a spot with an R_f similar to that of the red pigment. The red and green pigments were both subjected to electrophoresis, but were both found to be neutral.

It is apparent that the pigment of the green pepper is chlorophyll, which in the presence of light undergoes photolytic degradation to produce the long-chain alcohol phytol. This is supported by the lack of colour of the berries, while in the transition state between red and green. The phytol is photodimerized and undergoes dehydrogenation and dehydration to give the carotenoid pigment, lycopene, which is the principal pigment of the red pepper. The fact that the colour change is rapid once the pigments have been extracted from the berries indicates that the berries contain an inhibitor which tempers the photolysis.

The project has proved to be useful in that it has involved:

1. a study of plant growth from seed to berry,
2. a study of *Capsicum* berries,
3. chemical methods of obtaining natural products from plants in a fairly pure form,
4. a study of these products using modern techniques. It has been possible to relate chemical evidence to observable botanical events, and so to elucidate biochemical pathways.

The project could be carried out using the tomato plant, another member of the Solanaceae family, but *Capsicum annuum* is a novel and interesting subject for study.

Paper chromatography of monosaccharides

D. J. JEWELL

Mixtures of reducing sugars can be investigated by paper chromatography. Aqueous solutions of various sugars are applied to a rectangle of chromatography paper and the paper sewn into a cylinder (see Fig.). The solvent-system is butan-1-ol/ethanoic acid (acetic acid)/water (60 : 15 : 25). When the solvent has reached almost to the top of the paper, the paper is removed, unrolled and the solvent front marked; the paper is then dried. To locate the aldoses, the paper is sprayed with phenylammonium hydrogen benzene-1,2-dicarboxylate (aniline hydrogen phthalate) and heated for 5 minutes in an oven at 100 °C. The reducing sugars appear as brown zones. Phenylammonium hydrogen benzene-1,2-dicarboxylate (aniline hydrogen phthalate) reacts only with aldoses. To locate the ketoses, e.g. fructose, the chromatogram is sprayed with a solution of carbamide (urea) in trichloroacetic acid and heated as before. It is customary to express the chromatographic

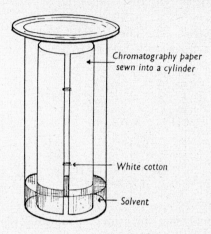

Chromatography paper
sewn into a cylinder

White cotton

Solvent

behaviour of sugars as R_o values, i.e. the ratio of the distance moved by a sugar to the distance moved in the same experiment by $2 : 3 : 4 : 6$ tetramethyl glucose.

Paper chromatography of lipids

D. J. JEWELL

A suitable lipid extract can be made by extracting 1 g of dried human blood plasma for 1 hour at room temperature with 5 cm³ of a mixture of trichloromethane/methanol (4 : 1). (Dried human plasma can usually be obtained from local hospital pathology laboratories. Plasma which is too old for medical use is quite suitable for this experiment.)

Discs of Whatman No. 1 filter paper (15 cm) are washed with the trichloromethane/methanol mixture and dried in air. A pencil dot is made in the centre of each paper, which is then placed on the rim of a Petri dish. Four or five drops of the lipid extract are applied to the centre of the paper drying the spot between each application. The diameter of the spot must not exceed 8 mm. The chromatogram is developed by applying the solvent drop by drop to the central spot, so that the final diameter reached is about 120 mm. Suitable solvents are methanol, benzene or propanone (acetone). The paper is again dried and cut into four sections for the following spot tests.

Unsaturated fatty acid residues (cis-octadec-9-enoic acid (oleic acid) and analogues) are tested for by immersing the paper in 1 per cent osmic acid for one minute. If positive, a brown-black colour appears.

Cholesterol is tested for by wetting the paper with a mixture of concentrated ethanoic (acetic) acid/concentrated sulphuric acid(VI) (1 : 1 v/v). A pink colour is positive.

Choline-containing lipids—e.g. phosphatidyl choline

$$
\begin{array}{l}
\quad\quad\ \ \overset{\displaystyle O}{\overset{\displaystyle \|}{}} \\
CH_2OC{-}R \\
\quad\ \ \overset{\displaystyle O}{\overset{\displaystyle \|}{}} \\
CHOC{-}R \\
\quad\ \ \overset{\displaystyle O}{\overset{\displaystyle \|}{}} \\
CH_2O{-}P{-}OCH_2CH_2\overset{+}{N}(CH_3)_3 \\
\quad\quad\ \ \overset{\displaystyle \|}{\underset{\displaystyle -}{O}}
\end{array}
$$

are tested for by immersing the paper in aqueous Reinecke salt [0.05 M solution of $NH_4(Cr\cdot(CNS)_4\cdot(NH_3)_3)$] for one hour. After the paper has been well washed with water, a pink colour shows the presence of choline-containing lipids.

The presence of *phospholipids* is shown by spraying the paper with a mixture of ammonium molybdate(VI) and chloric(VII) acid. It is then dried at 100 °C and exposed to hydrogen sulphide. A blue colour is positive. This test may be falsely positive if inorganic phosphate is present in the extract.

Paper chromatography of protein hydrolysates

D. J. JEWELL

Paper chromatography is the only satisfactory method for separating the amino acids resulting from the hydrolysis of proteins. For complete separation, 'two-way chromatography' is used. In this experiment, two proteins were used, gelatin and egg albumin. They were hydrolysed with 6 M hydrochloric acid at 100 °C in a sealed tube. This treatment is sufficiently mild to prevent total breakdown of tryptophan or tyrosine, both of which are sensitive to mineral acids. Alkaline hydrolyses were attempted but it was difficult to purify the resulting hydrolysates sufficiently for chromatography. The solvent-system used in the first separation was butan-1-ol/ethanoic acid (acetic acid)/water (40: 10: 50) and that used in the second was α-picoline/water/0.880 ammonia (70: 28: 2). The location reagent was 0·2 per cent ninhydrin in propanone. Just before use, a few drops of a 2 per cent pyridine solution are added to the reagent, which is then sprayed on to the paper. The propanone is allowed to evaporate and the paper is heated for 5 minutes at 105 °C. The amino acids react with the ninhydrin to give a purple or, sometimes, a brown colour.

Food colouring in Smarties
COLIN J. DODDS

In the Food Science topic of our CSE Mode 3 course, the pupils conclude their work by taking a quick look at some of the chemicals added to food. The object is to make them aware of the vast range of additives and our often incomplete knowledge of their long-term effects on the human body [1, 2].

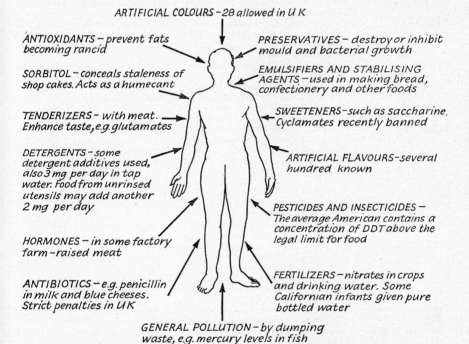

ARTIFICIAL COLOURS – 28 allowed in U K

ANTIOXIDANTS – prevent fats becoming rancid

SORBITOL – conceals staleness of shop cakes. Acts as a humecant

TENDERIZERS – with meat. Enhance taste, e.g. glutamates

DETERGENTS – some detergent additives used, also 3 mg per day in tap water. Food from unrinsed utensils may add another 2 mg per day

HORMONES – in some factory farm – raised meat

ANTIBIOTICS – e.g. penicillin in milk and blue cheeses. Strict penalties in UK

PRESERVATIVES – destroy or inhibit mould and bacterial growth

EMULSIFIERS AND STABILISING AGENTS – used in making bread, confectionery and other foods

SWEETENERS – such as saccharine. Cyclamates recently banned

ARTIFICIAL FLAVOURS – several hundred known

PESTICIDES AND INSECTICIDES – The average American contains a concentration of DDT above the legal limit for food

FERTILIZERS – nitrates in crops and drinking water. Some Californian infants given pure bottled water

GENERAL POLLUTION – by dumping waste, e.g. mercury levels in fish

The safety situation with regard to food dyes is far from clear. In 1957 the *Colouring matter in food regulations* banned all but thirty synthetic dyes. Since then two more have been withdrawn and other countries have also shortened their permitted lists. Even the UK and the USA cannot agree, for the USA has fifteen permitted food dyes, the UK twenty-eight, yet we ban nine of the USA dyes and many of ours are banned there.

TABLE I

The thirty food dyes permitted by the Colouring matter in food regulations, 1957

Common name of colour	Scientific name
Ponceau MX	disodium salt of 1-(2:4- or mixed-xylylazo)-2-naphthol-3:6-disulphonic acid
Ponceau 4R	trisodium salt of 1-(4-sulpho-1-naphthylazo)-2-naphthol-6:8-disulphonic acid
Carmoisine	disodium salt of 2-(4-sulpho-1-naphthylazo)-1-naphthol-4-sulphonic acid
Amaranth	trisodium salt of 1-(4-sulpho-1-naphthylazo)-2-naphthol-3:6-disulphonic acid
Red 10B	disodium salt of 8-amino-2-phenylazo-1-naphthol-3:6-disulphonic acid
Erythrosine BS	disodium or dipotassium salt of 2:4:5:7-tetraiodofluorescein
Red 2G	disodium salt of 8-acetamido-2-phenylazo-1-naphthol-3:6-disulphonic acid
Red 6B	disodium salt of 8-acetamido-2-*p*-acetamidophenylazo-1-naphthol-3:6-disulphonic acid
Red FB	disodium salt of 2-[4-(1-hydroxy-4-sulpho-2-naphthylazo)-3-sulphophenyl]-6-methylbenzothiazole
Ponceau SX	disodium salt of 2-(5-sulpho-2:4-xylyazo)-1-naphthol-4-sulphonic acid
Ponceau 3R	disodium salt of 1-*pseudo*cumylazo-2-naphthol-3:6-disulphonic acid
Fast Red E	disodium salt of 1-(4-sulpho-1-naphthylazo)-2-hydroxynaphthalene-6-sulphonic acid
Orange G	disodium salt of 1-phenylazo-2-naphthol-6:8-disulphonic acid
Orange RN	sodium salt of 1-phenylazo-2-naphthol-6-sulphonic acid
Oil yellow GG	a mixture of 4-phenylazoresorcinol and 4:6-di(phenylazo)resorcinol
Tartrazine	trisodium salt of 5-hydroxy-1-*p*-sulphophenyl-4-*p*-sulphophenylazo-pyrazole-3-carboxylic acid
Naphthol yellow S	disodium or dipotassium salt of 2:4-dinitro-1-naphthol-7-sulphonic acid
Yellow 2G	disodium salt of 1-(2:5-dichloro-4-sulphophenyl)-5-hydroxy-3-methyl-4-*p*-sulphophenylazopyrazole
Yellow RFS	disodium salt of 4-sulpho-4'-(sulphomethylamino) azobenzene
Yellow RY	disodium salt of 6-*p*-sulphophenylazoresorcinol-4-sulphonic acid
Sunset yellow FCF	disodium salt of 1-*p*-sulphophenylazo-2-naphthol-6-sulphonic acid
Oil yellow XP	3-methyl-1-phenyl-4-(2:4-xylylazo)-5-pyrazolone
Green S	sodium salt of di-(*p*-dimethylaminophenyl)-2-hydroxy-3:6-disulphonaphthylmethanol anhydride
Blue VRS	sodium salt of 4:4'-di(diethylamino)-4'':6''-disulphotriphenylmethanol anhydride
Indigo carmine	disodium salt of indigotin-5:5'-disulphonic acid
Violet BNP	sodium salt of 4:4'-di(dimethlylamino)-4''-di-(*p*-sulphobenzylamino)triphenylmethanol anhydride
Brown FK	a mixture consisting essentially of the disodium salt of 1:3-diamino-4:6-di(*p*-sulphophenylazo) benzene and the sodium salt of 2:4-diamino-5-(*p*-sulphophenylazo)-toluene
Chocolate brown FB	the product of coupling diazotized naphthionic acid with a mixture of morin and maclurin
Chocolate brown HT	disodium salt of 2:4 dihydroxy-3:5-di(4-sulpho-1-naphthylazo)benzyl alcohol
Black PN	tetrasodium salt of 8-acetamido-2-(7-sulpho-4-*p*-sulphophenylazo-1-naphthylazo)-1-naphthol-3:5-disulphonic acid

Sweets provide an ideal source for investigation and the wide variety of colours of Smarties makes them ideal. Half a dozen Smarties are required for each of the darker colours, or a dozen for the paler colours. The colour is extracted into 20–25 cm³ of warm water in a dish. The solution obtained contains the dyes and much sugar and will not give good chromatographic results unless the mixture is separated. It was found that quite a good separation could be obtained by using the solution to dye wool.

TABLE II

Dyes isolated from various coloured Smarties

Colour of Smartie	Dyes obtained								
	Brown	Yellow	Red	Carmine red	Blue	Mauve	Pink	Red orange	Orange
Yellow		√							
Orange								√	√
Red			√					√	√
Pink							√		
Violet					√	√	√		
Green		√			√				
Brown	√	√	√						
Black		√							

It is advisable to first prepare the wool, as even pure white wool usually contains some dyes, particularly fluorescing 'whiter than white' dyes. This may be extracted by boiling the wool with 1 per cent ammonia solution for four or five minutes and then rinsing the wool under the tap. This can if necessary be carried out by the laboratory technician before the lesson so that the wool can be used immediately.

TABLE III

R_f values for black Smarties with various solvents

Dyes	Solvent systems			
	Butan-1-ol 3 parts Ammonia 2 M 1 part Ethanol 1 part	Ethyl ethanoate 9 Glacial ethanoic 1 Water 1	Butan-1-ol Ethanol Ammonia 2 M Ethyl ethanoate	Ethyl ethanoate 1 Ethanol 1 Ammonia 2 M 3
Mauve	0.60	0.18	0.76	1.0
Blue	0.34	0.15	0.55	0.95
Carmine	0.25	0.70	0.46	0.70
Red	0.13	0.70	0.26	0.57
Yellow	0.09	0.12	0.24	0.78
Brown	0.04	—	0.10	—

Three to four feet of the prepared wool are placed in a 100 cm³ beaker and the dye extract added; the whole is acidified by adding dilute ethanoic acid.

The mixture is brought to the boil and simmered until most of the dye colour has been absorbed by the wool (about 4–5 min). The wool is then removed and rinsed thoroughly with cold water under the tap. The dye is re-extracted from the wool by simmering with 1 per cent aqueous ammonia for about five minutes, and then the aqueous extract is evaporated to dryness on a water bath. The residue is stirred with one drop of water and the solution spotted on a piece of Grade 1 chromatography paper. If each group in the class extracts a different colour, groups can share their extract and run several colours on their chromatogram. The chromatograms are run in gas jars or Shandon tanks using the ascending solvent technique; after trying a wide variety of solvent mixtures it was found that a butan-1-ol: ethanol: ammonia mixture (3 : 1 : 1) gave the best overall results.

Experience with this experiment invariably shows that the class are impressed by the rainbow array of six colours obtained from black Smarties (Table II). The experiment can readily be extended to identify the actual dyes (with more advanced courses) by comparison with known food dyes.

REFERENCES

1. Bicknell, F., *Chemicals in Food* (Faber, 1960).
2. Pyke, Magnus, *Synthetic Foods* (Murray, 1971).

Confirmation of sodium, potassium and magnesium by paper chromatography

D. AINLEY

DETERMINATION OF THE R_f VALUES OF SODIUM, POTASSIUM AND MAGNESIUM CHLORIDES

The chromatograms from which the mean R_f values of sodium, potassium and magnesium chlorides were determined, were prepared in a home-made chromatographic tank. Although these were descending chromatograms (i.e. the solvent eluted in a downward direction) and the later analyses were carried out on ascending chromatograms, it was found that there was no significant difference in the R_f values in the two types of chromatogram and the descending type enabled more precise measurements to be made. The tank, as can be seen from the accompanying photograph, consisted of a large glass jar fitted with a plastic screw cap from which was suspended by glass rods the bottom of a plastic detergent bottle to act as a solvent trough and a glass rod to act as an anti-siphon device. A length of Whatman No. 1 chromatographic paper strip (5 cm in width) was folded about 4 cm from one end to give a wick to fit into the trough and a pencil line was drawn across the paper about 2 cm from the fold to act as the origin line. The line was divided into four equal parts and spots of molar solutions of sodium, potassium and magnesium

chlorides were placed on the three dividing marks with a glass jet. The latter
was washed out with the solution by dipping into the solution and shaking
out several times and then, after dipping once again, it was shaken out gently
until one drop remained. On touching on to the paper a circular spot of
about half a centimetre in diameter was obtained. The jet was washed out
with distilled water between the addition of each solution.

Methanol (BDH Laboratory Reagent) was added to the solvent trough and
a small volume was added to the jar to minimize evaporation from the paper
during elution. The wick of the paper was threaded between the glass rods

into the solvent and, with the paper hanging freely, the cap was screwed
gently on to the jar. Elution was allowed to continue with the jar undisturbed
until the solvent front had travelled about 18 cm from the origin line and the
paper was then removed, the position of the solvent front being immediately
marked. The chromatogram was developed to render the spots visible by see-
sawing the paper through a 0.1 per cent solution (w/v) of dichlorofluorescein
in ethanol (*not* 50 per cent ethanol/water as Vogel suggests) and then, without
washing, through a 0.1 M solution of silver nitrate(v), the chlorides showing
up as bluish-red stains on a yellow or pink ground. The paper was then
washed with a stream of distilled water from a wash bottle and dried in the
dark on a sheet of chromatographic paper. The R_f values of the chlorides
were calculated by dividing the distance from the origin line to the centre of
the spots by the distance from the origin line to the solvent front.

The procedure was carried out several times and excellent agreement was obtained with mean values of 0.43 for sodium chloride, 0.24 for potassium chloride and 0.68 for magnesium chloride.

QUALITATIVE ANALYSIS OF SIMPLE SALTS

To approximately 0.1 g of the salt were added 10–15 drops of concentrated hydrochloric acid and the mixture was warmed. Water was then added drop-wise until a clear solution was obtained. This was spotted as in the previous determination on to a 1.5 cm strip of chromatographic paper and the paper was freely suspended in a corked boiling tube containing a few cm^3 of methanol (the spot must be above the surface of the solvent during elution), the strip being fastened to the underside of the cork with a drawing pin. Elution was allowed to proceed with the tube undisturbed until the solvent front had moved 10–12 cm and, after marking its position, the chromatogram was then developed, as before, in dichlorofluorescein solution and silver nitrate(v) solution. The R_f value of the chloride was calculated and compared with the mean values obtained in the previous determination.

When the method was tried out with a class of sixth-formers for the confirmation of the cations in three compounds A sodium hydrogen carbonate, B (magnesium sulphate(vi)) and C (potassium nitrate(v)) which were given as unknowns, the following values were obtained:

A 0.47, 0.41, 0.43, 0.48, 0.41, 0.41, 0.40
B 0.72, 0.70, 0.73, 0.69, 0.58, 0.73
C 0.25, 0.26, 0.25, 0.24, 0.25

showing excellent agreements with the mean values previously obtained.

Some trouble was encountered with 'tailed' or elongated spots—this appeared to arise when the paper touched the sides of the boiling tube and care must be taken to see that this does not happen. Another possible source of confusion in the case of magnesium compounds was the presence of excess hydrochloric acid in the test solution—hydrochloric acid has an R_f value in this type of chromatogram nearly the same as that of magnesium chloride and thus, if a magnesium compound is suspected, the quantity of hydrochloric acid should be reduced. In the case of a sodium or potassium compound two spots, one due to the alkali metal chloride and one due to hydrochloric acid, may show up on development but no difficulty should be encountered in differentiating between the two. Should a sodium compound be contaminated with the corresponding potassium compound, two spots again show up on the chromatogram with the relative sizes and depths of colour indicating which of the two cations is the impurity. The average time for the confirmation from start to finish was found to be about 40 min with the elution taking about 30 min.

REFERENCE

Vogel, *Macro and Semimicro Inorganic Analysis*.

Confirmation of the presence of chloride, bromide and iodide ions in qualitative analysis by paper chromatography

D. AINLEY

Preliminary work on the separation of halides, eluting with Vogel's solvent—90 per cent by volume of pyridine and 10 per cent of water—indicated that the R_f value of a halide depended not only on the anion but also on the cation. Thus, potassium, sodium and ammonium bromides were found to have the following R_f values:

potassium bromide	0.23
sodium bromide	0.43
ammonium bromide	0.56

The consequence of this was that, if a spot containing sodium chloride and potassium bromide was eluted with the above solvent, three spots showed up on development—one due to potassium chloride (R_f value 0.08), one due to potassium bromide and sodium chloride (R_f values 0.23) and one due to sodium bromide (R_f value 0.43). It was obvious, therefore, that, if the separation of halides on paper was to be used in the method of analysis described earlier, the halides must be applied in the form of salts containing a common cation and, since it is usual practice in confirming anions to eliminate heavy metal ions from the unknowns by boiling them with excess sodium carbonate solution, it was decided that the common cation should be the sodium ion and that the halides should be applied as sodium salts. The R_f values of the sodium halides were found to be:

sodium chloride	0.23
sodium bromide	0.43
sodium iodide	0.64

The method of analysis finally employed is summarized below.

PREPARATION OF THE SOLUTION FOR CHROMATOGRAPHIC ANALYSIS

To about 0.2 g of the compound or mixture were added 0.5–1 g of anhydrous sodium carbonate and about 5 cm³ of distilled water. The mixture was boiled until all the solid was dissolved and then filtered or centrifuged, if necessary. If an ammonium salt is present, ammonia will be evolved at this stage and the solution should be boiled until all this has been evolved—failure to do this will lead to additional confusing spots owing to ammonium halides appearing on the final chromatogram.

PREPARATION OF THE CHROMATOGRAM

The chromatograms were prepared in a home-made chromatographic tank, constructed from a screw-capped glass jar with the bottom of a plastic detergent bottle, acting as a solvent trough, suspended by glass rod from the cap (see preceding experiment). A length of Whatman No. 1 chromatographic paper strip (5 cm in width and about 25 cm in length) was folded about 4 cm from one end to give a wick to fit into the trough and a pencil line was drawn across the paper about 2 cm from the fold to act as the origin line. The line was divided into five equal portions and on the first three dividing marks were placed spots of molar solutions of sodium chloride, bromide and iodide, and on the fourth, the test solution prepared as in the previous section. Should the sodium halides be unavailable, the potassium salts added to and warmed with excess sodium carbonate solution will do equally well. The spots were added to the paper with the aid of a glass jet which was first washed out with the solution by dipping into the solution and shaking out several times, and then, after dipping once again, it was shaken out gently until one drop remained. On touching on to the paper, a circular spot was obtained. The jet was washed out with distilled water between the addition of each solution.

The solvent mixture consisting of 90 per cent by volume of pyridine (BDH Laboratory Reagent) and 10 per cent of water (the fact that most people find the smell of the reagent nauseating means that the work is best carried out in a fume cupboard) was added to the solvent trough and a small volume was added to the jar to minimize evaporation from the paper during elution. The wick of the paper was threaded between the glass rods into the solvent and, with the paper hanging freely, the cap was screwed gently on to the jar. Elution was allowed to continue with the jar undisturbed until the solvent front had travelled about 18 cm from the origin line and the paper was then removed, the position of the solvent front being immediately marked. The paper was then allowed to dry thoroughly before development—the presence of excess pyridine in the paper seems to inhibit development, particularly of chloride spots.

DEVELOPMENT OF THE CHROMATOGRAM

The dry paper was see-sawed through a 0.1 per cent solution (w/v) of dichlorofluorescein in ethanol and then, without washing, through a 0.1 M solution of silver nitrate(v). The paper was washed in a stream of tap water and dried in the dark by suspending the paper from a glass rod with a paper clip. The spots show up as light red or pink stains on a pale pink or light yellow ground.

INTERPRETATION OF THE CHROMATOGRAM

As has already been described, it was now only necessary to compare the final positions of the spots from the unknown with those from the known halides to make a reliable identification.

For class use where the number of 'tanks' may be limited or where the solvent mixture may be considered too expensive for large-scale use, four separate chromatograms from four different samples of unknown can be prepared on a single strip of paper in one tank and, instead of comparing the spots with standards, the R_f value of each spot is calculated. These are then compared with the values for the sodium halides listed above. The presence of other anions such as sulphate or nitrate does not appear to affect the R_f values of the sodium halides but, where potassium ions are present, secondary spots due to potassium halides may show up on the chromatogram. These will, however, be much smaller than those due to the sodium halides and should not cause confusion. The R_f value of sodium carbonate under the conditions of the experiment is zero and, if a large excess is used in preparing the test solution, the only effect will be to give a large spot on the origin line.

The total time of elution in these analyses was about 3 hours but, since the tank requires no attention during elution, it can remain undisturbed until the end of a school session or some suitable change of period when it requires only a minute to remove the paper and hang it up to dry. The chromatogram suffers no deterioration on prolonged drying in air, prior to development.

REFERENCES

Vogel, *Macro and Semimicro Inorganic Analysis*.

Identification of a Group 2A precipitate

D. AINLEY

The apparatus used in the elution process is the home-made chromatographic 'tank' (see page 26).

The completion of the analysis in the conventional way usually produces two or three crops of the group 2A (the copper group) precipitate; if this happens, one portion should be used for the conventional identification, while a second portion is used for chromatography. If only one crop is obtained, a solution of the precipitate is prepared and a few drops are retained for chromatography, the remainder being used for the conventional identification. The precipitate is dissolved in 1 : 1 concentrated nitric(v) acid–water and 1 : 1 concentrated hydrochloric acid–water, warming if necessary, to produce a solution which is approximately $M/5$ with respect to the cation(s).

A 20 cm length of 5 cm chromatographic paper strip is folded about 8 cm from one end to form a 'wick', which may be tapered so that it will fit more easily into the solvent trough, and a pencil line is drawn across the paper, about 2 cm from the fold, to act as the origin line. The line is divided into five equal portions by dots, placed at centimetre intervals, and on the first three dots are placed spots of $M/5$ solutions containing cations which may be identical to the cation(s) contained in the 'unknown'. For example, if the original precipitate is dark brown in colour, solutions of bismuth(III) nitrate(v) oxide (bismuth oxynitrate), copper(II) nitrate(v) and mercury(II) nitrate(v) should be used (see the Fig. below). If the precipitate is yellowish, spots of cadmium and mercury(II) nitrates(v) should be applied and, if lead appeared in group 1, one of the spots should consist of lead(II) nitrate(v) solution. On the fourth dot is placed a spot of the solution of the precipitate. The spots are again added by means of a glass jet, which is first washed with the solution as described previously and which should be washed with distilled water between each addition. The spot should not be too big—a spot about 5 mm in diameter is desirable.

The solvent which, in this case, consists of a mixture of ethanol (90 parts by volume), water (5 parts) and concentrated hydrochloric acid (5 parts), is added to the solvent trough of the 'tank' and a small volume is added to the jar to minimize evaporation during elution. The wick of the paper is threaded between the anti-siphon rod and the glass rod supporting the trough, so that the fold rests on the anti-siphon rod and the paper hangs vertically without

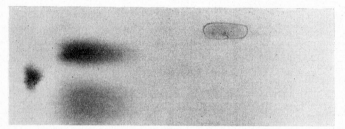

copper(II) ion

bismuth ion

mercury(II) ion

unknown ion

A dark brown group 2A precipitate (obtained in an A-level practical examination) is shown to be bismuth sulphide

touching the side of the trough, and the cap is screwed gently on to the jar. The paper should be cut to length at this stage, so that it does not touch the bottom of the jar or dip into the solvent in the jar. Elution is allowed to continue with the jar undisturbed until the solvent front has advanced to within one or two centimetres of the bottom of the paper, and the paper is then removed.

The chromatogram is treated to render the spots visible by holding the paper over yellow ammonium sulphide solution and, when the spots have become partially visible, by dipping the lower half into the solution and washing under the tap. The copper spot should not be dipped—it may completely disappear if it is so treated. The positions and colours of the spots from the unknown are now compared with those of the spots from the known solutions and a reliable identification is thus made.

The method is particularly useful when the original precipitate contains more than one cation and provides, for example, a suitable way of detecting cadmium in the presence of copper without the use of potassium cyanide solution. The time of elution is 3–4 hours but, since the tank requires no attention during this time, the paper can be placed in the tank before a school session and then removed at the end of the session or at a suitable change of period, with treatment of the chromatogram following at some later time.

Identification of a Group 4 precipitate

D. AINLEY

If the precipitate is thought to contain cobalt(II), nickel(II) or manganese(II) sulphide, the principle of the identification is the same as that used in the previous analysis. The precipitate is dissolved in a mixture of concentrated hydrochloric and concentrated nitric(v) acids and a spot of the solution is added, using a glass jet as before, to a 5 cm strip of chromatographic paper, alongside spots of M/5 solutions of cobalt(II) nitrate(v) and nickel(II) sulphate(vi) and a spot of a M solution of manganese(II) sulphate(v). Only a small volume

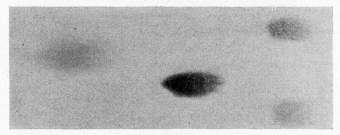

nickel(II) ion

cobalt(II) ion

manganese(II) ion

unknown ion

A black group 4 precipitate (obtained in the analysis of a mixture of ferrous sulphate and nickel carbonate) is shown to consist of nickel sulphide

of the solution is required for spotting and the remainder can be used for a conventional identification of the cation(s), thus causing no disruption of the normal scheme of analysis. The paper is eluted in a 'tank' with the solvent consisting of propanone (87 parts by volume), water (5 parts) and concentrated hydrochloric acid (8 parts) and the spots are made visible by dipping the paper in yellow ammonium sulphide solution and washing under the tap. The position of the spot from the unknown relative to those of the known compounds is then a reliable means of identification. The manganese spot may be difficult to see immediately after treatment with the locating agent but, on drying, it rapidly darkens and becomes clearly visible (see the Fig. above).

The cobalt(II) ions are visible as a blue spot during elution and elution should be allowed to continue until this spot has moved to within an inch of the bottom of the paper to give a good separation. The time of elution is 4–5 hours.

The separation works very well, even in the hands of a beginner. The only snag, encountered in the method, occurred when the atmosphere in the tank was not saturated with solvent vapour, thus permitting considerable evaporation from the surface of the paper and reducing the R_f value of the cobalt spot to about that of the manganese spot. This difficulty can be eliminated by placing the solvent in the trough and in the jar, some hours before the introduction of the paper. If the propanone, used in making up the solvent, is contaminated with water, the cobalt spot does not appear blue during elution, thus making it difficult to monitor the separation.

If the precipitate is thought to consist of zinc sulphide, it is dissolved, as before, in a mixture of concentrated hydrochloric and nitric(v) acids and a spot of the solution is eluted with the same solvent on a 5 cm paper strip, alongside a spot of $M/5$ zinc nitrate(v) solution, elution being allowed to continue until the solvent front has travelled to about an inch from the bottom of the strip. The paper is then removed and lightly sprayed with a solution of copper(II) sulphate(VI) (0.1 per cent w/v in distilled water containing 10 per cent v/v 1 M sulphuric(VI) acid) and then with a solution of mercury(II) chloride (2.7 g) and ammonium thiocyanate (3 g) in water (100 cm³). After about a minute, the zinc salts show up as pale mauve zones and, if two such zones appear, the original precipitate contains zinc ions.

If the precipitate is thought to contain zinc sulphide and another Group 4 sulphide, two chromatograms are prepared as described in the previous paragraphs, the spots on the first being located with ammonium sulphide solution and those on the second with the ammonium mercury(II) thiocyanate reagent.

Tables useful in the separation and identification of inorganic ions by paper chromatography

F. H. POUARD, J. F. W. McOMIE AND H. M. STEPHENS

Other ions may also be identified using the technique of paper chromatography. Here are some tables reproduced from *SSR*, 1952, **122** which the reader might find useful.

TABLE I

Examples of solvent mixtures used for certain separations

Solvent mixture	Group of cations
Butan-1-ol saturated with 3 M HCl	Pb^{2+}, Cu^{2+}, Bi^{3+}, Cd^{2+}, Hg^{2+}
Pentane 2,4-dione (acetyl acetone) saturated with water 0.5% v/v HCl and 25% v/v propanone (acetone)	As^{3+}, Sb^{3+}, Sn^{2+}
Glacial ethanoic acid (acetic acid) 25% v/v methanol	Fe^{2+}, Al^{3+}, Cr^{3+}

TABLE I (CONTINUED)

Solvent mixture	Group of cations
Propanone containing 5% v/v water and 8% v/v of HCl (d. 1.18 g cm³)	Ni^{2+}, Mn^{2+}, Co^{2+}, Zn^{2+}
Pyridine containing 20% v/v water and 1% w/w potassium thiocyanate	Ca^{2+}, Sr^{2+}, Ba^{2+}
Methanol	Li^+, Na^+, K^+
60% pentan-1-ol, 10% benzene, 30% conc. HCl	Al^{3+} from Fe^{2+} and Ti^{3+}
Conc. HCl and butan-1-ol (95 : 5) saturated with 1-chlorobutane	Sr^{2+} from Ba^{2+}, Ca^{2+} and Mg^{2+}
Mixture of methanol and butan-1-ol (4 : 1)	Na^+ and Li^+

TABLE II

Reagents used to identify cations on Chromatograms

Reagent	Cations
Potassium chromate(VI)	Pb^{2+}—yellow; Ag^+—brick red
Dithiooxamide $(CSNH_2)_2$ plus ammonia	Ni^{2+}—greyish blue; Cu^{2+}—olive green; Co^{2+}—brown
Sodium phosphinate (NaH_2PO_2) in 2 M HCl (warm)	As^{3+}—dark stain
Potassium thiocyanate and propanone (acetone)	Bi^{3+}—yellow; Fe^{2+}—red; Co^{2+}—blue; Cu^{2+}—black
Potassium hexacyanoferrate(II)	Fe^{2+}—blue
Rhodizonic acid, followed by HCl	Ba^{2+}—bright red; Pb^{2+}—violet
Zinc uranyl ethanoate (in 10% ethanoic acid)	Under ultra-violet light: Na^+—bluish-green fluorescence
8-Hydroxyquinoline (5 g) in 60% ethanol (1 litre)	Under ultra-violet light (over NH_3): Al^{3+}—green fluorescence (not quenched by ethanoic acid Sn^{2+}—yellow fluorescence (not quenched by ethanoic acid Cd^{2+}—yellow fluorescence (quenched by ethanoic acid Zn^{2+}—yellow fluorescence (not quenched by ethanoic acid Ca^{2+}—green fluorescence (quenched by ethanoic acid Mg^{2+}, Sr^{2+}, Ba^{2+}—bluish-green fluorescence (quenched by ethanoic acid) Most other cations give dark spots: Zn^{2+}—mauve (2–3 minutes)

TABLE III

Separation and identification of anions

Solvent: butan-1-ol (80), pyridine (40), water (80), ammonia (d 0.880).   Upper layer used as mobile phase

Anion	Procedure for detection	Colour of spot	R_f value
F^-	Lead ethanoate (acetate), wash strip in water, dry, hold over ammonium sulphide	Dark brown	0.12
Cl^-	Silver nitrate(v), wash strip in water, dry, expose to light or ammonium sulphide	Dark brown or black	0.15
Br^-	As for chloride ion	Dark brown or black (tails a little)	0.28
I^-	Lead(II) ethanoate	Yellow	0.50
ClO_3^- (i)	Saturated phenylammonium sulphate(VI) solution (10 cm³) and conc. sulphuric(VI) acid (10 cm³). Spray thoroughly	Blue	0.43
ClO_3^- (ii)	Potassium iodide in freshly prepared 2M HCl and warm	Brown	0.43
BrO_3^-	As for chlorate(v) ion (ii) but without warming	Brown	0.24
IO_3^-	As bromate(v) ion	Brown	0.08
NO_2^-	As bromate(v) ion	Brown	0.15
SO_4^{2-}	Aqueous mixture of barium nitrate(v) and pot. manganate(VII) followed by puffs of HCl gas	Pinkish-purple spot persisting 30 s after paper has been decolorized (an uncertain test)	−0.02
SO_3^{2-}	Pot. dichromate(VI) in 1 M sulphuric(VI) acid	White or pale green spot on an orange background	0.15
S^{2-}	Sodium pentacyanonitrosylferrate(II) (nitroprusside)	Cherry red (fades) (sulphide spot must be fresh)	0.15
PO_4^{3-}	Ammonium molybdate(VI) and nitric(v) acid (warm); follow with ammonium sulphide	Yellow blue	0.04
CO_3^{2-} (i)	Silver nitrate(v), wash in distilled water and warm to approx. 70–80 °C	Brown	0·11
CO_3^{2-} (ii)	Mixture of equal vols of water and universal indicator	Bluish-green (on pink background)	0.11
CH_3COO^-	As for carbonate ion (ii)	Yellow (on a pink background)	0.12
SCN^- (i)	Iron(III) chloride containing 2 M HCl	Reddish-brown	0.55

TABLE III (CONTINUED)

Anion	Procedure for detection	Colour of spot	R_f value
SCN$^-$ (ii)	Cobalt(II) nitrate(V) containing propanone (acetone) (warm)	Blue	0.55
AsO$_3$$^{3-}$	Concentrated copper(II) sulphate(VI) solution	Green	0.08
AsO$_4$$^{3-}$	Silver nitrate(V)	Brown	0.07
CrO$_4$$^{2-}$	As for arsenate(V) ion	Brick-red	0.05
(Fe(CN)$_6$)$^{3-}$	Iron(II) sulphate(VI) in 1 M sulphuric(VI) acid	Blue	0.08
(Fe(CN)$_6$)$^{4-}$ (i)	Iron(III) chloride in 2 M hydrochloric acid	Blue	0.07
(Fe(CN)$_6$)$^{4-}$ (ii)	Copper(II) sulphate(VI)	Brown	0.07

N.B. All reagents are of the usual concentration used for analysis, unless otherwise stated

Thin-layer chromatography

Preparation of chromatostrips and detection of components

D. A. STEPHENS

PREPARATION OF CHROMATOSTRIPS

The chromatostrips are prepared by dipping $15 \times 2\frac{1}{2}$ cm glass slides into a slurry of adsorbent in trichloromethane. When the slides are withdrawn, a thin uniform film of adsorbent adheres to them. It takes about three minutes for the trichloromethane to evaporate from the strips, which are then ready for use. If desired, the coated strips can be heated in an oven for 2–3 minutes to enhance the activity of the adsorbent, but this is not necessary for the experiments described. The normal microscope slide may be used to support the adsorbent, although the length of the adsorbent film is often too short to allow complete separation. 15 cm microscope slides, which are obtainable commercially, are preferable; we use window glass cut to size by the local plumber.

Trichloromethane has a number of advantages over water as a dispersion medium for the slurry. Water reacts rapidly and irreversibly with the binder contained in the adsorbent. Thus the slurry sets within two or three minutes, which means that it must be used quickly, and there is inevitably wastage of the adsorbent. The strips must then be heated in an oven for half an hour to remove all the water before they are ready for use. A slurry made with trichloromethane does not set and may be kept indefinitely if evaporation losses are made good. There is virtually no wastage of adsorbent when coating the strips and, as the trichloromethane evaporates rapidly, the strips are ready for use almost at once.

(a) Silica gel adsorbent

The adsorbent, Kieselgel G Nach Stahl, is obtainable from Camlab (Glass) Ltd, Cambridge. Make a slurry of silica gel (35 g) with trichloromethane (100 cm^3) in a stoppered flask. Transfer to a narrow cylindrical container, e.g. a salad-cream bottle or a 200×4 cm diameter gas jar. Hold two clean, grease-free microscope slides back to back and insert them into the well shaken slurry. Withdraw them at a steady speed, separate them and, holding them horizontally, wave them gently to and fro to accelerate the evaporation of the trichloromethane. If desired, up to one-third of the trichloromethane may be replaced by methanol which is cheaper.

(b) 'Polyfilla' adsorbent

'Polyfilla' is a cheap adsorbent which serves to illustrate the basic technique. It is, however, coarse and impure, so that a uniform slurry is difficult to prepare. It is a cellulose-based material which has a low adsorptive power, so R_f values bear no relation to those obtained on silica gel, and the sample spots tend to be large.

Thoroughly grind Polyfilla in a mortar and sieve it through a fine mesh gauze. To the sieved Polyfilla (20 g) in a bottle or a small gas jar, add trichloromethane (50 cm³) slowly, stirring all the time. Keep the container stoppered. For some work, satisfactory results can be obtained without sieving.

COMPILER'S NOTE

R. Shirley (SSR, 160, 767) suggests that dental quality calcium sulphate(VI) may be used as an absorbent. A weak suspension in distilled water proved more successful than more volatile solvents which made large-grained layers and the slides could be dipped in by hand and removed horizontally. No spreading mechanism was needed since most of the water ran off leaving a layer held to the slide by surface tension. This natural phenomenon ensured evenness of layer across the slide. When almost dry the slides can be baked in an oven at 80–100 °C.

APPLICATION AND ELUTION OF THE SAMPLE

After wiping clean the rear surface and edges of the slide, apply the sample with a capillary tube. Capillary tubes of a suitable size are prepared by drawing out melting point tubes, although other glass tubing will do. Spot about 1 microlitre of a 1 per cent solution on to the adsorbent about 6 mm from the short edge of the strip. It is important not to punch a hole in the adsorbent during the process and to keep the size of the spot small. This is the most difficult part of the technique and requires some practice.

Allow the solvent to evaporate and place the slide in a jar containing a 2–3 mm depth of eluting solvent. Coffee jars are a convenient shape. To obtain reproducible results, the vapour in the jar should be saturated with solvent vapour; hence the jar should be closed and lined with filter paper soaked in the eluant. If the lining is omitted, the results are not reproducible, but the resolution is better. The sample spot must be above the level of the eluting solvent or it will be washed off the strip. When the solvent has reached the top of the strip (35 minutes or less, depending on the solvent) or at any time after 15 minutes, withdraw the slide and allow the eluant to evaporate.

DETECTION OF COMPONENTS

It is advisable in the introductory stages to use materials which are visible, as the nature of the process is then readily apparent. Most compounds are not normally visible so the adsorbent must be sprayed with a reagent which will react with the components to give a coloured derivative or one which will fluoresce in ultra-violet light. Sprays driven by compressed air or a hand bellows are obtainable and a useful spray gun incorporating a cartridge of propellant is now marketed. Spray reagents are also available in aerosol packs but they are expensive.

Iodine vapour is a useful detecting agent and has the advantage that it does not permanently affect the compounds. The strip is placed in a tank or bottle containing a few crystals of iodine. If the bottle has a metal top, it should be covered with cardboard or the iodine will corrode it. The components show up as brown spots and the development is rapid—about 5 minutes.

Corrosive sprays such as concentrated sulphuric(VI) acid may be used. If, after spraying, the slides are placed in an oven, most organic substances are carbonized and become visible as black spots.

Preservation of specimens

J. M. FREEMAN AND S. A. SEDDON

One of the apparent disadvantages of thin-layer chromatography is that the student cannot retain the chromatogram on the microscope slide and insert this in his laboratory notebook.

However, it is possible to retain the chromatogram by using the following technique.

A piece of clear 'Fablon' is cut out measuring 3 cm by 8 cm. The paper backing is peeled off and the clear material is placed on a piece of glass, sticky side facing upwards. The slide is placed carefully, chromatogram downwards, on the Fablon. The forefinger is run over the back of the slide, gently at first, and then fairly firmly.

The slide is turned over and the pressing with the forefinger repeated over the whole surface for 15 seconds. The Fablon is lifted at a corner gently and peeled off the microscope slide. The chromatogram will have adhered to the underside of the Fablon.

The piece of Fablon is then inserted in the appropriate place in the student's notebook.

By a development of this technique it is possible that colours which fade on exposure to air, e.g. the ninhydrin amino acid complexes, can be retained. If such is the case this is an additional advantage of t.l.c. over paper chromatography.

To preserve colours which fade on prolonged contact with air, proceed as above. When the chromatogram has been removed from the slide by the Fablon, lay it on the piece of glass, sticky side facing up.

Cut out a piece of Fablon 2.5 cm by 7.5 cm and lay this, sticky side downwards, on the piece of Fablon in such a way that there is a sticky border left on the lower piece of Fablon. Press the top piece of Fablon on to the lower piece and run the forefinger over the surface as before.

Stick this in the student's notebook.

This method has been used with amino acids, but the length of time the chromatogram was left was not long enough for complete confirmation of the success of the method.

Spreading and preparation of large plates

W. M. M. BARON

Take about a dozen clean photographic plates and lay them side by side, long sides touching, on a piece of plastic, covering a flat, even area of bench. Clamp a ruler at one end (see Fig.) so that the plates cannot slip. Then stick 'Sellotape' along the two outside edges of the plates. This serves two func-

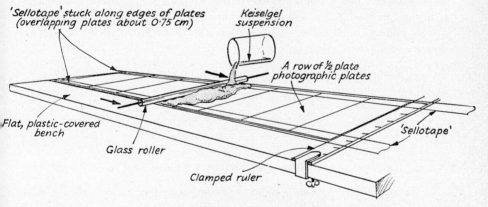

Spreading thin-layer chromatogram plates

tions; it prevents the plates from slipping and also keeps the roller at exactly the right height above the plate, so that a thin layer of Kieselgel is obtained.

Construct a roller from a wide, strong glass tube (25 cm long, 1 cm bore) together with a 30 cm long solid glass rod, slid down the centre of the tube. Place this at the unclamped end of the plates. An assistant should then pour a good quantity of the Kieselgel mixture on to the plate in front of the roller. Move the roller evenly and steadily towards the clamp. *Only move the roller once across the plates.* The assistant must take care to keep enough Kieselgel in front of the roller. Leave the plates on the bench for about one hour to dry, remove the 'Sellotape' and then bake them in an oven at between 80° and

100 °C for about one hour. The plates may then be kept in a desiccator for
a few days until they are needed. There is an indication that plates kept for
more than a week may show some deterioration.

Before the plates are used any traces of 'Sellotape' should be carefully
cleaned off (using a clean rag and a little ethanol) and also a 1 cm band of
Kieselgel from the edges of the plate should be scraped off. This will prevent
a 'billowing' of the solvent as it comes into contact with the edge of the plate.
Handle the plate at the edges only; never touch the Kieselgel.

Solvents that can be used for elution

W. M. M. BARON

The R_f value of a substance, i.e.

$$\frac{\text{distance moved by substance}}{\text{distance moved by solvent front}}$$

depends principally upon the nature of the eluant and the nature of the
sorbent layer. For a given adsorbent the R_f value increases with increasing
solubility in the eluant. Substances normally dissolve well in solvents of
similar polarity, e.g. salt in water and paraffin wax in benzene. The order of
polarity of solvents commonly used is:

Light petroleum	1,2-Dichloroethane
Cyclohexane	Propanone (acetone)
Tetrachloromethane	Ethanol
Benzene	Methanol
Trichloromethane	Water
Ethoxyethane (diethylether)	Pyridine
Ethyl ethanoate (ethyl acetate)	Ethanoic acid (acetic acid)

If the nature of the components to be separated is known, then a likely
solvent can be chosen. It is common practice to use mixtures of solvents
whose proportions can be adjusted so that the components are distributed
over the length of the chromatogram. If the nature of the components is not
known, then a suitable solvent can be found by testing a variety of solvents.

Solvent effect in thin-layer chromatography

D. R. BROWNING

The following experiment shows, in startling fashion, how the developing
solvent in thin-layer chromatography affects not only the R_f values of a

compound, but in some cases the order of elution of the components of a mixture.

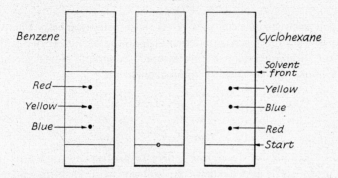

Two glass plates (5 × 15 cm) are coated with 'Polyfilla' (see page 39) and allowed to dry. A drop of the mixture of dyes (see Table) is placed on each plate as shown in the diagram and allowed to dry.

The plates are placed in benzene and cyclohexane as the elution solvents and the following results obtained (distance travelled by the solvent front = 15¼ cm):

Compound	R_f (benzene)	R_f (cyclohexane)
Indophenol (Blue)	0.23	0.51
p-Dimethylaminoazobenzene (Yellow)	0.51	0.80
Sudan Red G (Red)	0.79	0.24

R_f values in thin-layer chromatography are not repeatable and the values given above can only be taken as an approximate measure of the amount of separation obtainable. It can be seen that the order of elution of Sudan Red G (red spot) is completely reversed.

The separation of ink pigments

D. A. STEPHENS

Concentrate a sample of ink by evaporation and chromatograph in butanol/ethanol/2 M ammonia (3 : 1 : 1). A good separation takes about 20 minutes. Better results are obtained if a narrow felt-tipped pen is touched directly on to the adsorbent layer. The most interesting results are obtained with black ink which contains several dyes, none of which is black.

Separation of cis- and trans-isomers of azobenzene

D. A. STEPHENS

Azobenzene consists of a mixture of cis- and trans-isomers. The proportion of the cis-isomer in the mixture depends upon the recent history of the sample because the interconversion is phototropic. A sample stored in the dark for about six hours will be entirely trans-azobenzene.

$$
\begin{array}{ccc}
\text{Ph—N} & \xrightleftharpoons{h\nu} & \text{Ph—N} \\
\| & & \| \\
\text{N—Ph} & & \text{Ph—N} \\
(\text{trans}) & & (\text{cis})
\end{array}
$$

Prepare a sample by dissolving 0.05 g of azobenzene in 5 cm³ petroleum ether (b.pt 80–100°) and irradiate the solution with an ultra-violet lamp for 20 minutes. Chromatograph one microlitre of this sample and a sample of a similar solution which has been kept in the dark side by side using benzene as the eluant. The cis-isomer has the lower R_f value.

This experiment illustrates one of the advantages of chromatography: namely, the simple separation of closely similar substances that would require considerable time and effort by conventional methods.

Separation of the indicators in Universal Indicator

D. A. STEPHENS

Reduce about 20 cm³ of the Indicator as supplied by BDH Ltd, to about 1 cm³ and chromatograph a sample of the concentrate in propanone (acetone) methanol (1 : 1). The indicators are rendered more easily visible by exposure of the adsorbent to ammonia vapour, but the colours fade. A permanent record is obtained if the strip is sprayed with 2 M sodium hydroxide.

To identify the constituent indicators, note the colours of the spots under acid and alkaline conditions, select likely indicators from a standard text and chromatograph them alongside the Universal Indicator on one strip or a series of strips. With some skill, three spots can be placed on one strip. Complete correspondence between the standard indicator and a component in the mixture strongly suggests that they are the same compound.

This exercise illustrates a standard method of identification of the components of a mixture. By considering the additive colours of the component indicators and the pH's at which they change, the continuous variation of the colour of Universal Indicator with pH can be explained.

Monitoring the esterification of benzoic acid

D. A. STEPHENS

This standard A-level preparation involves a period of refluxing for 3 hours. During this time the class may be usefully occupied in following the progress of the reaction by means of thin-layer chromatography.

Take samples of the initial reaction mixture and the reaction mixture after various times, e.g. $\frac{1}{2}$, $1\frac{1}{2}$, and 3 hours, and chromatograph them in benzene/methanol (30:1). Spray with 0.2 per cent rhodamine B in ethanol and view by ultra-violet light. Benzoic acid and ethyl benzoate are visible as magenta spots on a brighter yellowish-pink background. As the experiment progresses, the size of the ethyl benzoate spot increases while that of the benzoic acid decreases. Benzoic acid has the lower R_f value.

Alternatively, expose the chromatograms to iodine vapour. This is a more convenient, but less sensitive method of detection.

The identification of amino acids in a mixture

D. A. STEPHENS

Hydrolysis of a protein will yield fifteen or more different amino acids. The resolution of so complex a mixture of closely related compounds was almost impossible until the advent of paper chromatography, which enabled the analysis to be completed in a matter of days. With thin-layer chromatography the time required is a few hours. Glass plates are needed if a complete separation is to be achieved, and they require a sophisticated spreader to coat them with slurry. With microscope slides, the technique can be illustrated if a simple mixture of amino acid is used.

Make up a mixture of lysine, cysteic acid, proline, glycine, valine and leucine in 0.1 M aqueous hydrochloric acid/ethanol (1:1). Spot a sample of the mixture together with reference sample of any of the amino acids individually, e.g. glycine and leucine. Elute with butan-1-ol/ethanoic (acetic) acid/water (3:1:1). Dry the strip in an oven and spray with ninhydrin reagent. On heating, five purple spots and one yellow spot (proline) appear. The R_f values are in the order—leucine > valine > proline > cysteic acid > lysine.

Ideally, reference spots should not be needed as the R_f value is a characteristic of a compound, just as melting point is. Under these conditions, however, the reproduceability of the R_f values is insufficiently high.

Ninhydrin reagent may be purchased in aerosol sprays ready for use from BDH Ltd. Alternatively the solution is made up as follows: Dissolve ninhydrin (1 g) in butan-1-ol (400 cm^3 and ethanoic (acetic) acid (100 cm^3); dilute the solution with an equal volume of ethanol before use.

Separation of inorganic ions

D. A. STEPHENS

Thin-layer chromatography may be applied to all the problems to which paper chromatography has been applied, and the following experiment shows an application to qualitative inorganic analysis.

Spot a sample of a 1 M solution of Co^{2+}, Mn^{2+}, and Ni^{2+} on to a strip. Develop the chromatogram in propanone (acetone)/ethoxyethane (diethyl ether)/6 M HCl (50 : 50 : 1). Allow the solvent to evaporate, expose the strip to ammonia vapour and spray it with 1 per cent butanedione dioxime (dimethylglyoxime) in ethanol. The nickel complex is red, the others are yellowish brown. The R_f values are in the order $Fe^{3+} > Co^{2+} > Mn^{2+} > Ni^{2+}$. The Fe^{3+} is not included in the sample as it is a contaminant in the silica gel and forms a band at the front of the finished chromatogram.

Cobalt may be visible before and during development as a blue spot.

Separation of phenols

D. A. STEPHENS

Place a spot of a solution containing benzene-1,3,5-triol, benzene-1,3-diol, 3-nitrophenol and 2-nitrophenol (3 per cent of each in ethanol) on the base line of two separate strips. On one strip put spots of 2-nitrophenol and 3-nitrophenol; on the other strip put spots of benzene-1,3-diol and benzene-1,3,5-triol. Elute with methyl benzene/dioxan (6 : 1). Dry the strips in an oven at 110° for 2/3 minutes and spray them with dichlorfluorescein.

The phenols are visible to varying degrees in daylight and are readily distinguishable as purple spots when viewed by ultra-violet light. The most difficult spot to detect is 3-nitrophenol and a higher concentration than that indicated may be required. R_f values are in the order 2-nitrophenol > 3-nitrophenol > benzene-1,3-diol > benzene-1,3,5-triol.

The effect of acid on a condensation reaction of propanone (acetone)

D. AINLEY

The Oxford and Cambridge Special Paper in 1966 included a question on the effect of low pH values on the rate of the reaction between propanone and phenylhydrazine. This does not appear easy to demonstrate but the use of thin-layer chromatography enables the effect of acid on the reaction between propanone and 2,4-dinitro-phenylhydrazine to be simply shown. The reaction is carried out in solution in benzene and the investigation is based on the fact that propanone 2,4-dinitro-phenylhydrazone can be separated from the

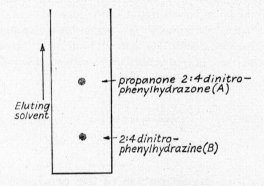

unchanged dinitro-phenylhydrazine by placing a spot of a solution containing the two on a plate coated with silica gel and gypsum and eluting with benzene. The resulting chromatogram appears as in the figure above. The compounds show up as yellow spots and the disappearance of the spot B indicates that the reaction is complete. The plates are $15 \times 2\frac{1}{2}$ cm glass strips, coated as described on page 38.

A cold solution of 2,4-dinitro-phenylhydrazine in benzene is prepared by adding about a gramme of the solid to a boiling tube nearly full of the solvent, stoppering the tube and shaking thoroughly. When the excess crystals have settled, two 10 cm³ portions of the solution are decanted into two separate boiling tubes and to the first portion is added one drop of concentrated hydrochloric acid by means of a dropping pipette, the mixture being thoroughly shaken. The formation of a solid at this or a later stage can be disregarded. A spot of the solution is then placed 1–2 cm from the bottom of the coated plate using a capillary tube or glass jet—if the depth of colour of the spot when the solvent evaporates is not sufficient, a second and a third spot can be placed on top of the first, the solvent being blown away after each addition. Two drops of propanone are then added from a dropping pipette and, after shaking, a spot of the solution is placed alongside the first on the silica gel layer. Further two-drop portions of propanone are added and a spot of the mixture after each addition is spotted on the same or a second plate. A completed plate is eluted in a screw-capped jar containing about ½ cm depth of benzene, elution being allowed to proceed until the solvent has travelled about 4 cm beyond the line of spots. The reaction appears to be complete after the addition of 4 to 6 drops of propanone but, if the procedure is repeated with the second 10 cm³ portion of the dinitro-phenylhydrazine solution, this time omitting the addition of acid, the mixture formed by the addition of a dozen drops of propanone still contains unreacted dinitro-phenylhydrazine.

The effect is even more marked when one drop of concentrated sulphuric(VI) acid is added to 10 cm³ of dinitro-phenylhydrazine solution, 3 drops of propanone being apparently sufficient to cause complete reaction. This could be due to the lower concentration of hydrogen ions in the solution when sulphuric acid is used. Peter Sykes reports that the maximum rate of reaction occurs at pH 4, the rate decreasing as the pH falls, and the sulphuric(VI) acid here could provide the required acidity without starting to inhibit the reaction as might occur with hydrochloric acid.

REFERENCES

Sykes, Peter, *A Guidebook to Mechanism in Organic Chemistry* (Longmans, 1961), 145.

Chromatographic investigation of the products of nitration of phenol

D. AINLEY

The technique of thin-layer chromatography provides a simple but elegant method for the identification of the products formed by the action of dilute nitric(V) acid on phenol, the results making a useful starting point for a discussion on the directive effect, in electrophilic substitution, of a hydroxyl group attached to a benzene ring, and the relative ease of the attack on phenol by an electrophilic reagent in such substitution.

The nitration of phenol and the confirmation of the presence of 2- and 4-nitrophenol in the product forms part of Experiment 13.3b—the nitration of arenes—in the Nuffield A-level course but in that experiment the identities of the products are supplied to the pupil and merely confirmed. In the following procedure, only the possibility of unchanged phenol in the reaction mixture is suggested to the pupil, the mono-nitrophenols being identified without further direction.

NITRATION OF THE PHENOL

1 cm³ of a 90 per cent solution of phenol in water is added to a test-tube and to this solution is added, in small portions, dilute nitric(V) acid, prepared by

THIN-LAYER CHROMATOGRAPHY 49

adding 4 cm³ of water to 1.5 cm³ of concentrated nitric(v) acid. The mixture should be stirred well and addition should be made at such a rate that the temperature is kept within the range 45–55 °C. When the addition is complete, the mixture is shaken or stirred for a few minutes and is then allowed to stand for about 10 minutes. The product is now transferred to a separating funnel, 50 cm³ of benzene are added and the mixture is thoroughly shaken. After being allowed to settle, the lower aqueous layer is discarded and the benzene extract is dried over magnesium sulphate(vi).

IDENTIFICATION OF THE PRODUCTS

The identification involves the use of 15 × 2.5 cm glass plates coated with Silica Gel G, using the method described on page 38, a pair of plates, held in contact, being dipped in a freshly shaken slurry of Silica Gel G (35 g) in trichloromethane (100 cm³) and then separated and allowed to dry. The backs and edges of the plates are cleaned of adsorbent before use. Spots of solution are added to the plates using small jets, made by drawing out melting point tubes in the flame of a micro burner.

To one of the plates are added spots of the benzene extract and a dilute solution of 2-nitrophenol in benzene, the spots being placed in line, about 1 cm apart and about 1¼ cm from the bottom of the layer. The plate is then eluted with benzene in a screw-capped jar until the solvent front has risen to within a few mm of the top of the layer. The solvent is allowed to evaporate from the plate and a comparison of the positions of the spots will confirm the presence of 2-nitrophenol in the products of nitration.

On a second plate are placed, in line, spots of the benzene extract, a dilute solution of 3-nitrophenol in benzene and a similar solution of 4-nitrophenol and the plate is eluted with benzene as before. This time, however, when the benzene has evaporated, the plate is eluted a second time with a mixture of benzene (3 vol) and ether (1 vol) and, after elution, the spots are located by standing the plate for a few minutes in a screw-capped jar containing iodine crystals. A comparison of the positions of the spots will now confirm the presence of 4-nitrophenol and the absence of the meta-derivative.

It is likely that, amongst the spots produced by the mixture after the second elution, will be one, slightly above that of the meta-derivative and not in line with any of the spots of the known mono-nitro-compounds. This can be shown to be a spot of unchanged phenol by adding a spot of the benzene extract and a spot of a dilute solution of phenol in benzene to a third plate and eluting the plate, first with benzene and then with the benzene/ethoxy-ethane (diethylether)mixture. The spots are again located in the iodine bottle.

The second elution causes the separation from the benzene extract of spots of compounds other than the mono-nitro-derivatives, including a purple compound which could be a quinone derivative, produced by the oxidation of the phenol. The identification of the other products in the mixture could make an interesting exercise.

The nitration of phenol

M. HARDY

The use of thin-layer chromatography to separate and identify the isomeric products, as was stated, is a useful introduction to discussion of the directive effect of the phenolic hydroxyl group. In addition, steam distillation of the reaction mixture, followed by t.l.c. examination of the residue, steam distillate and a little of the original mixture, provides an excellent demonstration of the efficiency of the steam distillation as a separation technique. The presence of only 2-nitrophenol in the steam distillate leads to discussion of the significance of hydrogen bonding in this means of separation.

For the separation of phenolic products a mixture of hexane and methyl-benzene (toluene) (1 : 6) is very useful with silica gel as the stationary phase adsorbent. Development of the chromatogram by spraying with dichloro-fluorescein followed by u.v. irradiation shows the compounds as dark spots on a bright background, illustrating the strong absorption of the aromatic ring in the u.v. region.

An investigation of the optimum reaction conditions for nitration provides an excellent project. Direct nitration involving aqueous nitric(v) acid invari-ably causes formation of tarry by-products. Using glacial ethanoic (acetic) acid as solvent, concentrated nitric(v) acid does not suffer from this drawback, which might be expected from the use of this nitration mixture on easily oxidized heterocyclic compounds such as furan.

There is some evidence that nitric(III) acid (nitrous acid) is involved in the mechanism of nitration. This may be removed from nitric(v) acid by boiling with urea but the concentration of the latter appears to be critical since heating concentrated acid with 1–2 per cent by weight of urea does not eliminate the formation of by-products, whilst using 5–8 per cent urea renders the acid completely ineffective as a nitrating agent.

T.L.C. of chlorophyll pigments

W. M. M. BARON

The pigments may be extracted by grinding the leaf in pure propanone (acetone). It is probably best to purify the extract (see page 16) but good results can be obtained without such purification. Do not dry the spots with a heater when you apply them.

Solvents

(a) 100 parts petroleum ether (60–80 °C)
    20 parts propanone (pure)

(b) 100 parts petroleum ether (100 °C)
   12 parts propanone (pure)

No developing spray is needed, but the spots fade and should be marked immediately.

R_f values

		Solvent (a)	Solvent (b)
Carotene	yellow	0.96	0.90
Phaeophytin	grey	0.44	0.28
Xanthophylls (?)	yellow	0.35 & 0.19	0.22
Chlorophyll a	blue-green	0.23	0.11
Chlorophyll b	yellow-green	0.17	0.10
Faint green-yellow spot			0.08

T.L.C. of plant acids (after Blundstone 1963)
W. M. M. BARON

The acids are probably best extracted from plant material by grinding in 70 per cent ethanol. The solution can then be made more concentrated by gentle evaporation.

Solvent

   10 parts butyl methanoate*
   4 parts 98 per cent methanoic acid
   1 part water
   with sodium methanoate (0.05 g in 100-cm³ solvent)
      Solid bromophenol blue (added until a pale orange colour is obtained)

         * take care with these reagents.

Dry the chromatogram carefully and then develop it by holding over a bottle of ammonium hydroxide. Do not let the chromatogram get too blue. Mark the yellow acid spots that appear against the violet blue background. The spots sometimes become clearer if the chromatogram is left for some days.

R_f values (*calculated to the bromophenol blue front*)

2,3-Dihydroxybutane-1,4-dioic acid (tartaric acid)	0.32
2-Hydroxypropane-1,2,3-tricarboxylic acid (citric acid)	0.50
2-Hydroxybutanedioic acid (malic acid)	0.61
2-Oxopropanoic acid (pyruvic acid)	0.85
Butanedioic acid (succinic acid)	0.92

REFERENCE

Blundstone, H. A. W., 'Paper Chromatography of Organic Acids', *Nature*, **26**, January 1963, 377.

T.L.C. of amino acids

W. M. M. BARON

Solvents

(*a*) Phenol saturated with water (see page 14)

(*b*) 120 parts butan-1-ol
30 parts glacial ethanoic (acetic) acid
60 parts water

These chromatograms are here run only one way, for two-way chromatography a second solvent must be run at right angles to (*a*) or (*b*) above.
A useful solvent is

(*c*) 180 parts ethanol
10 parts ammonia solution
10 parts water

The chromatogram must be developed by spraying with ninhydrin (0.2 g in 100 cm^3 butan-1-ol; a useful aerosol is available from BDH) and the plate heated strongly for the spots to appear.

R$_f$ *values*

		Solvent (*a*)	Solvent (*b*)
Phenyl alanine	brown	0.52	nil
Valine	red-purple	0.41	0.39
Cystine	pink	0.32	0.26
Tyrosine	purple	—	0.25
Arginine	deep red-purple	0.32	0.23
Glycine	orange-red	0.29	0.20

Plant extracts (e.g. potato sap) give a good range of amino acids, a particularly good separation being obtained with phenol solvent (*a*).

T.L.C. of sugars

W. M. M. BARON

Solvents

Solvent (*b*) under amino acids, above. Phenol (solvent *a*) would probably also be satisfactory.

The sugar should be extracted and applied in 0.1 M sodium ethanoate (acetate) solution. The chromatogram is developed by spraying with a strong solution of 4-methoxyphenylammonium chloride (*p*-anisidine hydrochloride) in butan-1-ol (1.5 g in 50 cm³). A drop or two of concentrated hydrochloric acid may be added to this reagent, which must be freshly prepared. Heat the plate strongly for about three minutes for the yellow-brown spots to appear.

R_f values seem rather variable, but the following have been obtained with solvent (*b*) (butan-1-ol/ethanoic (acetic) acid/water):

Glucose	0.43
Fructose	0.31
Sucrose	0.24

REFERENCE

Baron, W. M. M., *Organization in Plants* (Arnold, 1963).

Isolation of organochlorine pesticide residues from carrots

T. P. BEAUMONT, JANE JESSOP, JANE RAPER AND JOANNA STEPHENSON

PROCEDURE

Carrot seed (Scarlet Horn) was sown at the beginning of April in virgin soil in a suitably divided plot. At the beginning of July the plants were sprayed as follows.

Plot 1	DDT
Plot 2	BHC
Plot 3	DDT + BHC
Plot 4	Control, unsprayed

The foliage was saturated with spray made at quadruple strength. Protective clothing (overall, gloves, face-shield) was worn and care taken to shield rows not being sprayed, from spray drift.

The carrots were harvested in September, washed, chopped into small cubes and reduced to a coarse pulp in the 'liquidizer' of a food mixer. The pulp was filtered at the pump using a Buchner funnel with fine holes, and without filter paper. The solid residue was macerated with propanone (acetone), allowed to stand overnight and again filtered. The two filtrates were mixed, made up to about 1 dm³ with water, and about 25 g of sodium sulphate(VI) were added. This was extracted with two 100 cm³ portions of 60–80° petroleum ether. Vigorous shaking was found to lead to the formation of emulsions which only slowly separated into two layers. The combined petrol extracts were dried over anhydrous sodium sulphate(VI) and the petrol was then distilled off over a water bath until about 5 cm³ remained. This was transferred to a smaller flask and the volume further reduced to about 0.1 cm³. This concentrated solution was subjected to chromatography.

Rubber gloves were worn throughout the extraction process.

THIN-LAYER CHROMATOGRAPHY

Glass slides, 15 × 2½ cm, were coated with Kieselgel G (nach Stahl) by dipping in pairs into a slurry of the adsorbent in chloroform and dried in the air. Approximately 2 'microlitre' spots of the extracts were applied by means of capillary tubes and the plates run in hexane.

The developing agent was a 0.5 per cent ethanolic silver nitrate(v) solution and a 1 per cent fluorescein solution mixed in equal volumes just before use. A Humbrol Jet-Pak proved a satisfactory spray. The spots were visualized under a soft u.v. lamp, the pesticides appearing as light spots. Strongly fluorescing spots were obtained at R_f values of 0.80–0.90. These were assumed to be due to pigments or other materials from the carrots, since no attempt had been made to 'clean up' the extract.

The DDT proved more difficult to detect than the BHC. Superior visualizing agents are mentioned in the monograph (1). These were not tried.

The results were as follows:

	R_f ($\times 100$) ranges
Plot 1 (DDT)	38–45
Standard DDT	32–35
Literature value	42
Plot 2 (BHC)	16–18
Standard BHC	17–18
Literature value	21

Less satisfactory results were obtained from the plot 3 extract (sprayed with DDT and BHC). Only a small quantity of material was available and was spotted on to an unsatisfactory batch of plates. The R_f values obtained did not correspond with the others. However, when spots of standard mixtures were run on plates deliberately allowed to deteriorate, similar R_f's were obtained to those of the plot 3 extract.

No spots located around R_f's 0.2 or 0.4 were obtained from the extract for plot 4 (unsprayed control).

The chromatography was done over a period of several weeks, and care had to be taken to avoid inconsistent R_f values resulting from two main factors:

1. The method of coating the plates does not permit of accurate control of layer thickness. Unusually thick layers gave poor results. A batch of plates can be checked to ensure the thickness is within acceptable limits by running a spot of standard mixture (BHC and DDT) and checking the R_f values.

2. Even though the bottle containing the slurry was kept tightly sealed when not in use, some deterioration of the adsorbent occurred slowly. This

was manifested by a rise in the R_f values as the layer lost its activity. It was found that re-activation of the plates in an oven at 120 °C corrected this condition.

PAPER CHROMATOGRAPHY

The extracts and standard substances were subjected to reversed-phase paper chromatography using Whatman No. 3 MM coated with liquid paraffin BP (applied as a 10 per cent v/v solution in ethoxyethane (ether) and a 70 per cent v/v solution of propanone in water. The BHC was readily detected in extracts from plots 2 and 3, giving spots corresponding to the quoted R_f value of 0.66. The DDT proved more elusive, possibly because the recommended visualizing agent (phenoxyethanol and silver nitrate(v)) was not available, but eventually a clear spot was obtained from the plot 1 extract in good agreement with the literature value of 0.39.

POSSIBLE FUTURE DEVELOPMENT

No attempt was made to assess the quantity of pesticide present. This would undoubtedly prove difficult, but an able sixth-former might be able to make some estimate, especially if several plots sprayed with mixtures of DDT and BHC of varying composition were employed.

REFERENCE

Thompson, J. and Abbot, D. C., *Pesticide Residues* (R.I.C. Lecture series, 1966).

Column chromatography

Preparation of the column

T. I. WILLIAMS

The apparatus required is extremely simple and available in any school laboratory. The figure shows a typical apparatus for chromatographic analysis. Percolation may take place under gravity alone, or may be made more rapid, either by applying suction to the Büchner flask or pressure to the top of the column through a rubber bung fitted with a bicycle valve. The

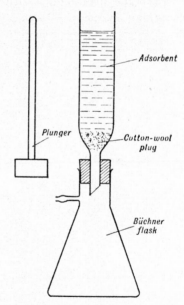

latter alternative is preferable, since reduction of pressure often disturbs the column by formation of air-bubbles. The size of the column can vary enormously according to the amount of material to be dealt with. For class use a convenient size is 2.5 × 12 cm. The containing glass tube should be 5 to 10 cm longer than this to provide a reservoir for solution or solvent.

Perhaps the most difficult part of the technique is the packing of the column with the adsorbent. Irregular packing is reflected in irregularly shaped bands. For alumina one of the best methods is as follows: The bottom of the tube is plugged with a small cork; it is then filled three-quarters full with the solvent used in making the solution for analysis. When bubbles have stopped rising from the cotton-wool plug (see Fig.), alumina is added gradually and sinks to the bottom of the tube. When filled, the tube is allowed to stand for half an hour or so to allow the column to settle. The cork is then removed from the lower end and the tube is allowed to drain. When almost all the solvent has run through, the solution to be examined is run on to the top of the column, great care being taken to avoid disturbance of the surface of the adsorbent. When once wetted the column must not be allowed to become dry until development is complete, as this causes cracks to appear and the adsorbent may shrink away from the glass. Alternatively the column may be packed dry, small quantities of adsorbent being added at a time and pressed down by tapping gently with a plunger. The latter need be nothing more elaborate than a cork stuck on the end of a glass rod (see Fig.). The diameter of the plunger should be only slightly less than that of the column itself. Columns that have been packed dry need not be wetted with solvent before adding the solution to be examined.

Extrusion

When development is judged sufficient, the column is sucked or drained dry. A stiff rod is then introduced through the lower end of the tube, pressing against the cotton-wool plug. If the tube is of uniform bore, and well packed, the column of adsorbent should slide out quite easily. If not, each zone may be scraped out with a spatula into a separate vessel (a Petri dish is very suitable as its rim prevents spilling of the adsorbent). As a rule, the boundaries of the zones are not quite straight, but with a little patience they can be closely followed if a fine spatula is used.

Experiments for class use

T. I. WILLIAMS

Experiment 1

Take 5 cm³ of a 0.1 per cent aqueous solution containing equal parts by weight of methylene blue and malachite green. On percolation the methylene blue is adsorbed as a sharply defined band at the top of the column, from which it can subsequently be eluted with ethanol, while the malachite green is completely washed through by thorough development with distilled water. Investigate the behaviour of other mixtures of dyes.

Experiment 2

5 cm³ of a molar solution of iron(III) alum and copper(II) sulphate are chromatographed on a similar column and developed by means of 50 cm³ of

distilled water. This gives a chromatogram brown at the top (iron), with a pale blue band below (copper). On further development with a 1 per cent solution of potassium cyanoferrate(II) the top band becomes blue (Prussian blue) and the lower brown (copper cyanoferrate(II)).

Experiment 3

5 cm³ of a solution of the nitrates(v) of lead, silver, zinc and cadmium are chromatographed and developed first with distilled water and then with ammonium sulphide solution. The appearance of the chromatogram is then:

Top:	Black	lead
	Grey	silver
	White	zinc
Bottom:	Yellow	cadmium

Possible applications to inorganic qualitative analysis should be obvious.

Experiment 4

Prepare a 40 per cent solution of butter in benzene (filter). Pass 5 cm³ through a column and develop with benzene. Compare with a chromatogram of margarine prepared under similar conditions.

For further experiments the following plant extracts should prove interesting, especially if they can be examined in ultra-violet light.

Plant	Organ	Absorbent	Solvent
Carrot	Root	Alumina or magnesia	Petrol ether
Maize	Grain	Alumina	Benzene + petrol
Orange	Peel	Calcium carbonate	Carbon disulphide
Dandelion	Petals	,, ,,	Petrol ether
Fucus vesiculosus	Fronds	Alumina	,, ,,

Test-tube chromatography and ion-exchange
KEITH I. P. ADAMSON

One of the disadvantages of column chromatography as a class experiment is the large volumes of eluant required. These will doubtless be collected, but their purification provides an extra job for a sixth-former, the laboratory steward or an already overworked chemistry teacher!

I have been able to effect considerable economies in this direction by using another laboratory 'waste product'. Third forms seem adept at pushing holes in the bottom of any test-tubes they are attempting to clean. I collect these and use them for experiments in column chromatography.

A wad of glass wool supports the column which is prepared from a slurry in the usual way. Although the column is much shorter than usual, separations on it are quite effective. Ink pigments and flower pigments have been successfully separated on such a column.

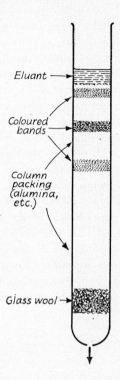

In this school two such columns are kept prepared for demonstrating ion-exchange processes. Packed with Zeocarb 225 and Deacidite FF, the colour changes of copper(II) dichromate(VI) are easily seen. The copper(II) ions are strongly adsorbed, but dichromate(VI) ions prove more persistent, so a dilute solution is to be preferred.

An ion-exchange column has also been used for the volumetric determination of metal ions in solution. Zeocarb 225 was first converted to the hydrogen form by washing 100 cm³ 5 M hydrochloric acid through the column. This was then rinsed with distilled water until free of acid. 25 cm³ portions of the test solution were run through this column and collected in conical flasks. Being now an acid, its concentration was determined by titration against standard alkali. Although it can be determined by other methods, copper(II) sulphate(VI) was used as providing a visual check on progress. Using 0.1 M solutions four titrations were performed without any sign of saturation of the test-tube column.

Preparation of 2,4-dinitrophenylhydrazones

D. HODSON

Recrystallization of 2,4-DNPH's from methanol is often attended by not inconsiderable loss in material, and frequently the product obtained is not of the desired degree of purity. Simple chromatography on an activated alumina column, using trichloromethane as eluant, affords the hydrazone in high purity. The most suitable alumina is type 'H', 100/200's mesh. A suitable chromatography column is readily constructed using a 50 cm³ burette, fitted with a plug of glass wool.

The crude 2,4-DNPH is dissolved in the minimum volume of trichloromethane, introduced on to the column and eluted with chloroform. The 2,4-DNPH passes quickly through the column whilst any unreacted 2,4-dinitrophenylhydrazine is strongly adsorbed and remains at the top of the column as a dark red band. In view of the immobility of this band in trichloromethane, quite large quantities of 2,4-DNPHs may be chromatographed on a single column. (I have successfully purified quantities as great as 9.5 g using a single alumina column measuring 32 cm × 0.5 cm.)

The eluate should be filtered, to remove any alumina that has been washed through the glass wool plug, before the solvent is removed under reduced pressure. Should a greater degree of purity be desired, the procedure may be repeated using a clean column.

'Dry column' chromatography

G. J. GREEN

A simplified version of the technique of 'dry column' chromatography is very suitable for the illustration of column chromatography in that it requires much less solvent than the conventional technique, and also all the compounds from the separation may be isolated within about half an hour of starting the procedure. The technique has been described in detail [1, 2], and usually involves deactivation of the adsorbent so that the results of t.l.c. may be reproduced on the column. For school use this deactivation procedure is not necessary, and the method may be used in the following way.

A length of thin-walled nylon tube [3] is heat-sealed at one end, and a plug of glass wool is placed at that end. Several small holes are made below the plug to prevent the formation of air pockets during packing of the column. Dry adsorbent is poured into the column (one-fifth of the adsorbent at a time) and the column is compacted by dropping on to a hard surface several times. When the column is filled and properly compacted it is quite rigid enough to be clamped to a retort stand. The mixture for separation is introduced at the top of the column and allowed to soak in. Development of the column is effected by addition of solvent from a dropping funnel. When the solvent front reaches the bottom of the column development is complete, and the column may be sliced with a knife to separate the components of the mixture. The segments of column are extracted with ethoxyethane (diethyl ether) or methanol or dichloromethane to obtain the pure components. As nylon is transparent to ultra-violet light (short wavelength) colourless compounds may be detected on the column.

The volume of solvent used for a 50 × 1 cm column is about 20–30 cm³, and so it may be seen that this technique uses much less solvent than the conventional method of 'wet column' chromatography. If just the required volume of solvent is placed in the dropping funnel the column may be left to develop and stop automatically. This enables a class to start the columns, continue with other work, and then return to slice them up and extract the components, rather than spending the whole time collecting fractions. This technique therefore not only enables the class to examine quite complex mixtures quickly, but also reduces the volumes of solvents used and saves the time of recovery of vast quantities of solvent.

REFERENCES

1. Loev and Snader, *Chem. and Ind.*, 1965, 15.
2. Goodman and Loev, *Chem. and Ind.*, 1967, 2026.
3. Available as 'C' gauge nylon tubing supplied in 100 and 1000 foot lengths flat diameters 1 in, 1½ in, 2 in, 3 in, from Walter Coles & Co. Ltd, Plastic Works, 47/49 Tanner St, London, S.E.1.

Chromatographic separation of yellow plastid pigments

D. MORGAN

STAGE I

Qualitative. Obtain fresh plant material for the extraction. Lettuce, cabbage, carrot and turnip tops are quite suitable. Tear the leaves and immerse in propanone (acetone)–petrol ether (60 : 40 mixture). Leave for two minutes, then macerate the tissues in a mortar.

Quantitative. Select the leaf material by careful sampling, then weigh the sample(s) carefully to 0.05 g and proceed as above.

STAGE II

Decant the supernatant extract of pigments, etc., and pour the mixture through a cotton or glass wool plug via a glass funnel into a separatory funnel. In quantitative estimation the macerated material requires further extraction with solvent, until the plant residue is colourless.

Wash the petrol ether fraction free from propanone with water. The ether layer will take up all the undissociated plastid pigments (chlorophylls, carotenoids, etc.).

After washing, the ether extract is freed of water by drying over anhydrous sodium sulphate(VI) for five minutes.

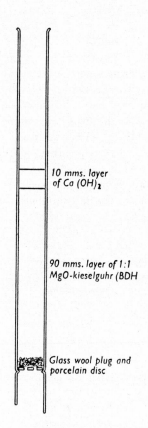

10 mms. layer of $Ca(OH)_2$

90 mms. layer of 1:1 MgO-kieselguhr (BDH

Glass wool plug and porcelain disc

Adsorption column used for the separation of leaf pigments

STAGE III

The chromatograph adsorption tube (see illustration) is prepared for the pigment fractionation. The glass tube should be packed dry. There are two types of adsorption mixtures:

(*a*) magnesium oxide : kieselguhr mixture (1 : 1) to about 9 cm. Pure calcium hydroxide layer to about 1 cm is placed above the MgO–kieselguhr.

(*b*) aluminium oxide : sodium sulphate mixture (1 : 1) to about 12 cm. The Al_2O_3 must be pure and the Na_2SO_4 must be preheated at 140 °C for

10 hours. The mixture as a whole functions best if it has been left for about four days after preparation.

A glass rod with a disked end is most suitable for preparing the column.

STAGE IV

Either type of prepared column is fixed to a Büchner filtering flask. The water pump is turned on and petrol ether is poured on to the top of each column set up. This solvent will be drawn through the column (under reduced air pressure). From now onwards the column must be kept wet with solvent.

The dried petrol ether extract of pigments can now be poured on to the column and drawn into the column very slowly. This first stage of adsorption is crucial and the more perfectly it is carried out the better will further stages in development be obtained.

STAGE V

After complete penetration of the extract, the column may be washed with about 10 cm³ petrol ether.

(a) Add about 25 cm³ propanone : petrol ether (10 : 90) in small amounts, very close to the top of the lime layer.

(b) In Al_2O_3–Na_2SO_4 columns, the chlorophylls may spread down the column for about 2 cm.

Development of the column is carried out as follows:

(a) Add about 25 cm³ propanone : petrol ether (10 : 90) in small amounts, drawing the solvent into the column. A further yellow band will move out of the green zone and eventually lie at the interface of the two adsorbent layers. At the same time the lower orange-pink zone will be washed out of the column. This eluate is the CAROTENE FRACTION.

The lime layer is now scraped off the lower column, using a spatula. It is discarded. The column is now washed with about 40–50 cm³ propanone : petrol ether (60 : 40), drawing the solvent mixture into the column. This treatment eventually washes the 'xanthophylls' out of the adsorbent. The eluate is the CAROTENOL FRACTION.

(b) In the alumina–sodium sulphate(VI) column the pigments are eluted by washing the column with about 25–30 cm³ propanone : petrol ether (40–60° b.pt) (1.2 : 98.8).

The 'xanthophylls' can be separated from the carotenes by fractionating the eluate (after washing it free from propanone with 4-hydroxy-4-methylpentan-2-one (diacetonol).

If a quantitative estimate of pigments is required, then the MgO : kieselguhr column will give excellent results. The eluates are again freed from propanone by washing with water, after which they are dried over anhydrous sodium sulphate(VI). The solutions can be made to standard volume and samples can be matched against predetermined standards in either a photoelectric cell/microammeter circuit or in a colorimeter.

Molecular sieve chromatography

G. A. MIERAS

Chromatography can be defined as 'the uniform percolation of a fluid through a column of more or less finely divided substance which selectively retards, by whatever means, certain components of the fluid' [1]. Provided 'finely divided substance' can be considered to include paper, this is a convenient general explanation of the process and as such covers paper chromatography, thin-layer chromatography and column chromatography.

One example of the latter type is molecular sieve chromatography which has been extensively developed in the last few years and is now a widely used technique. Like many scientific advances it was originally discovered by chance—during an attempted electrophoretic separation of amines on a column of starch in 1953 (in Sweden) someone on leaving the laboratory one night forgot to switch on the current. But despite this, on arrival the next morning it was observed that a certain degree of fractionation had taken place. The significance of this separation was immediately realized and a series of experiments carried out to try and capitalize on this discovery and develop the technique further. By coincidence the same observation was made almost simultaneously in two other research centres, one in France and one in America.

The experiments showed that starch was not a very successful column packing material [2] and the search therefore started for other materials working on the same principle. The major breakthrough came when a granular cross-linked dextran was tried [3] which absorbed water to form a gel. This material was found to be far superior to all others tried and soon after was marketed as 'Sephadex', a range of materials being produced. Gel filtration, as it was called at that time, had started. The term 'molecular sieve chromatography' was introduced later when the somewhat complex quantitative theories of the process were worked out [4] and is now the most generally used name.

Descriptively the process is very simple and is illustrated in the diagram below [5]. The gel swells in a solvent (usually water) to give particles containing a three-dimensional network of polymer chains, and these separate molecules according to the size of the molecules (although shape can be a contributing factor) as shown in the figure. The swollen gel particles have a porous structure with the pores being of such a dimension as to exclude particles above a certain size (known as the exclusion limit). When a solution of large and small molecules is loaded on to a column (*a*), molecules larger than the exclusion limit of the gel cannot enter the gel particles and instead pass through the liquid spaces surrounding the gel particles (known as the void volume of the column, V_o). Smaller molecules can enter the gel particles (*b*) and can therefore flow freely through the total volume of the liquid in the column, that outside the gel particles (V_o) plus that inside the gel particles (V_i). Obviously the availability of this much greater volume of liquid to the

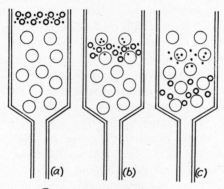

(*a*)   (*b*)   (*c*)

○   Swollen gel particles
o   Large molecules
•   Small molecules

smaller molecules means that they will be retarded in comparison to the larger molecules, i.e. they will travel through the column much more slowly than the larger—hence the separation of large and small molecules (*c*). It is also apparent that no molecules can be eluted from the column before a volume of liquid equal to the void volume has passed through the column, and similarly all will be eluted by the time the total column volume ($V_o + V_i$) of liquid has passed through the column.

The main uses of columns are desalting (i.e. the separation of molecules of greatly differing size) and fractionation (i.e. separation of molecules not differing so much in size). Columns can also be used in an analytical fashion by making use of the linear correlation between $\log_{10}$ molecular weight and the distribution coefficient of a solute,

$$K = (V_e - V_o)/(V_i - V_o)$$

(where V_e is the observed elution volume of a fraction from the column; V_o the void volume and V_i the internal volume as before). V_o is found by using a molecule large enough to be totally excluded from the gel particles—blue dextran (molecular weight about 1 000 000) is often used as it has a strong blue colour, and V_i is found using a very small molecular weight substance (e.g. the yellow dinitrophenyl derivative of an amino acid). The column can be calibrated using macromolecules of well substantiated molecular weight. Using the resultant calibration plot the molecular weights of unknown macromolecules can be estimated. The accuracy of the method (ca 10 per cent) is about as good as any of the other available methods of macromolecule molecular weight determination. By employing continual spectrophotometric analysis of column eluant related to careful volume measurement (weighed or drop-counted tubes in a fraction collector) this can easily be done. If the u.v. adsorption at 253 nm is used (characteristic adsorption of all protein-aceous material) colourless samples can also be identified in this way.

There are several types of column packing material available. Sephadex [6] gels are cross-linked dextrans which fractionate according to their degree of cross linking and hence porosity, e.g. Sephadex G-10 is the most highly cross linked and least porous and fractionates in the molecular weight range 0–700, whilst Sephadex G-200 is the most highly cross-linked and least porous and fractionates in the molecular weight range 5000 to 800 000. Also available are Bio-Gel [7] gels which are granulated polyacrylamide gels of varying composition also leading to differing amounts of cross-linking, and hence they display similar properties. There are also gels in both ranges which allow ion-exchange at the same time as fractionation; coarse and fine grades of gel which are selected depending on flow rate and degree of resolution required in the particular experiment; gels suitable for use in polar solvents; gels containing a certain amount of agarose in which the polymer chains are held together by hydrogen bonds instead of cross-linking—this leads to much wider fractionation ranges (e.g. Bio-Gel A-150m fractionates in the range 1.10^6 to 150.10^6). All these fractionation ranges are only approximate, of course, as both the shape and chemical structure of the molecules being separated as well as molecular size can have an effect on fractionation.

In addition to the uses already described many more have recently been introduced and are being introduced. An important example of these is the study of interacting systems—both relatively slow and fast reversible equilibrium systems can be detected qualitatively and to a certain extent quantitatively by this technique (e.g. associating–dissociating systems as in many proteins). Minimal molecular weights in such systems can also be estimated when the technique is employed in conjunction with other methods of molecular weight determination. One of the main values of the technique is sensitivity—very low concentrations of material can be used especially if spectrophotometric analysis is applied in conjunction with the technique.

In practice the technique of molecular sieve chromatography is fairly straightforward once the vagaries of column packing and loading have been mastered. It is a relatively speedy, efficient and accurate technique widely used in chemistry, biological sciences and medical research laboratories amongst others—anywhere concerned with macromolecule work is likely to use it. For the teacher it can no longer be sufficient to consider chromatography as the separation of ink on filter paper or the demonstration of a thin-layer chromatography plate. Column chromatography is rapidly superseding both of these in many areas and as it is not difficult in theory or practice when considered qualitatively (school kits are now available) it should certainly be mentioned when the topic of chromatography is discussed.

REFERENCES

1. Martin, A. J. P., *Ann. Rev. Biochem.*, 1950, **19**, 517.
2. Lindquist, B. and Storgards, T., *Nature*, 1955, **175**, 511.
3. Porath, J. and Flodin, P., *Nature*, 1959, **183**, 1657.
4. Anderson, D. M. W. and Stoddart, J. F., *Lab. Pract.*, 1967, **16**, 7, 841.
5. Male, C. A., *Lab. Pract.*, 1967, **16**, 7, 847.
6. Sephadex: 'Sephadex in gel filtration', *Pharmacia*, 1963, Uppsala, Sweden.
7. Bio-Gel: *Bio. Rad.*, 1966, Richmond, California, USA.

Gel permeation chromatography

J. H. JOHN PEET

Gel permeation chromatography (GPC) became established as a chromatographic technique in 1963 as a result of the work by J. C. Moore [1]. It is now widely used as a technique in organic, biological and inorganic chemistry. A variety of names have been given to this technique (including gel filtration, gel chromatography, molecular sieve chromatography, molecular exclusion chromatography and restricted diffusion chromatography). No one name has been universally accepted, though the use of the word 'gel' implies a limitation on the range of materials used. In fact, porous rigid materials such as glasses and silicas have been used. Useful reviews have been written by Heitz and Kern [2] and Cazes [3]. Several books are now available on the subject, and the new edition of Abbott and Andrews' book on chromatography [4] includes a description of the technique.

NATURE OF THE STATIONARY PHASE

The two most widely used gels are those distributed under the names Sephadex and Bio-Gel. Sephadex gels are prepared as spherical beads by cross-linking dextrans with 3-chloro-1,2-epoxypropane (epichlorohydrin) in alkaline solution. Bio-Gels are obtained by the polymerization of pro-penoamide (acrylamide) with hepta-2,5-diene-1,7-diamide (methylenebis (acrylamide)). The large number of hydroxyl groups gives the hydrophilic gels a great affinity for water and so they swell up in electrolyte solutions. The pore size is influenced by the extent of the cross-linking, since, as the degree of cross-linking is increased, the swelling is reduced.

If the hydroxyl groups are replaced by other suitable groups, hydrophobic gels can be obtained which can be used in non-aqueous solvents. If Sephadex is reacted with isocyanates, urethane-type gels are formed. Methyl 2-methylpropenoate polymerized in the presence of ethylene glycol dimethacrylate produces another hydrophobic gel, as does the co-polymerization of phenylethene (styrene) with 2 per cent diethenyl benzene (divinylbenzene).

The gels are classified by their *water regain values*, which are defined as the volumes (in cm³) of water taken up by 1 g dry gel grains. The gels are numbered such that the code number is ten times the *water regain value*. For example, Sephadex G-10 has a water regain value of one unit. The smaller values, therefore, have smaller pores. Sephadex G-10 can be used to separate molecules of up to 700 units molecular weight, and G-200 is used for molecules with molecular weights in the order of hundreds of thousands. The polysaccharide agarose (Sepharose) takes up even larger molecules.

The effectiveness of the separation is determined by the grades of each gel. The coarser grades are only suitable for easily resolved mixtures, see Table I.

TABLE I

Characteristics of some Sephadex resins

Grade	Particle diam. (10^{-6} m)	Molecular weight fractionation range for protein
G-25 Coarse	100–300	
G-25 Medium	50–150	1000–5000
G-25 Fine	20–80	
G-25 Superfine	10–40	
G-50 Coarse	100–300	
G-50 Medium	50–150	1500–30 000
G-50 Fine	20–80	
G-50 Superfine	10–40	

MECHANISM OF ACTION

No one theory appears to be adequate in explaining the observed effects of gel permeation chromatography. The main proposals can be grouped into four categories.

1. *Steric exclusion mechanism* [5]

There are a range of pore sizes in a gel bed, and the larger molecules 'find' the smallest number of pores. The molecules that are larger than the pore size are said to have exceeded the *exclusion limit*, and pass through the bed in the solvent phase only.

2. *Thermodynamic control* [6]

An equilibrium is set up between the molecules dissolved by the solvent in the pores and the solution outside the gel particles. This partition coefficient is affected by the molar volume of the solute and by the different chain conformations within the pores.

3. *Restricted diffusion mechanism*

The residence time of solute molecules in the gel pores is greater than the entry time. According to this approach the separation is diffusion controlled rather than an equilibrium process. Consequently the retention volume is affected by the flow rate. This process is dominant at high flow rates.

4. *Secondary exclusion effect*

This mechanism becomes effective in the overloading situation. The small molecules diffuse into the pores the most rapidly, thereby excluding the larger ones.

APPLICATIONS OF GEL PERMEATION CHROMATOGRAPHY

Obviously this technique can be used for the preparative or analytical separation of mixtures, and has been widely used in this field by biochemists. The technique has particular value in the separation of molecules of similar polarity but of different size, traditional chromatographic methods being less effective in these circumstances (Fig. 1).

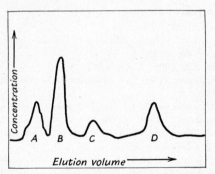

Fig. 1. *Separation of oligoethanediols on a polystyrene gel using THF as solvent:*
A. $HO-(-CH_2CH_2O-)-_{36}H$ C. $HO-(-CH_2CH_2O-)-_{18}H$
B. $HO-(-CH_2CH_2O-)-_{27}H$ D. $HO-(-CH_2CH_2O-)-_9H$

It became apparent early on in the investigation of GPC that there was some correlation between elution volumes and molecular weights. Heitz and Kern [2] showed that for a number of proteins this volume is a linear function of the logarithm of molecular weights (Fig. 2).

Some workers have used this technique to analyse the constituents of equilibrium mixtures, and others to measure reaction rates [7].

Brook [8] has used GPC to separate the geometric isomers fumaric and maleic acids, and other workers [9] have extended the procedure to the separation of inorganic salts.

Another development has been the use of thin-layer gel filtration [5], the superfine grade of resin being essential.

While a lot of research has been done into GPC, few experiments have been developed specifically for education purposes [10]. The makers of Sephadex

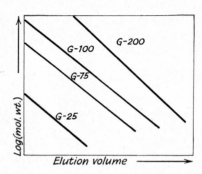

Fig. 2. Relationship of molecular weights to elution volumes

have produced such a kit, but there is room for a lot of project work by school and college students in this field.

REFERENCES

1. Moore, J. C., *J. Polymer Sci.*, 1964, A2, 835.
   Idem, ibid., 1965, C8, 233.
2. Heitz, W. and Kern, W., *Angew. Makromol. Chem.*, 1967, 12, **1,** 150.
   Idem, ibid., 1966, 8, **43,** A625.
3. Cazes, J., *J. Chem. Ed.*, 1966, 7, **43,** A567.
4. Abbott, D. and Andrews, R. S., *An Introduction to Chromatography*, 2nd edition (Longmans, 1970).
5. *Sephadex—Gel Filtration in Theory and Practice*, Pharmacia (GB) Ltd.
6. Hjertén, S., *J. Chromatog.*, 1970, 2, **50,** 189.
7. Ackers, G. K. and Thompson, T. E., *Proc. nat. Acad. Sci.*, 1965, **53,** 342.
8. Brook, A. J. W., *J. Chromatog.*, 1969, 3, **39,** 328.
9. Pecsok, R. L. and Saunders, D., *J. Chromatog.*, 1969, 3, **41,** 429.
10. Dewhurst, F., *J. Chem. Ed.*, 1969, 12, **46,** 864.

Gas chromatography

Some simple experiments using the principles of gas chromatography

D. R. BROWNING

The simple apparatus described below gives semi-quantitative results in a number of separations. The dimensions of the column are not critical for simple work. The column packing is 'Tide' detergent powder. (Mesh size >60.)

EXPERIMENTS

1. A small piece of capillary tubing is placed on the end of the column as in the diagram (*a*) and hydrogen gas passed for some time to sweep out all the

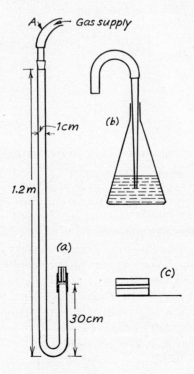

air. The hydrogen emitted from the column is lit (care must be taken in doing this). A mixture of carbon disulphide and benzene is injected into the system at point A using a syringe. (The carbon disulphide–benzene mixture must be kept away from the flame.) In a few minutes the hydrogen flame turns blue, then sooty, owing to the elution of carbon disulphide followed by benzene. A number of experiments are carried out with varying percentage compositions of the two substances and the corresponding variation in flame height noted. From the results, semi-quantitative estimations of relative concentration can be obtained.

2. In this experiment the capillary is replaced by the titration apparatus shown in (b) and the carrier gas is nitrogen. The injection system (c) consists of a short length of fine wire fastened to a piece of capillary tubing by 'Sello-tape'. This is used to inject a mixture of diethylamine and butylamine into the system at A. As the amines are eluted from the column into the water in the flask they are titrated with 0.01 M/HCl hydrochloric acid. A plot of titration against time (measured from the moment of injection) is drawn, and an integral chromatogram obtained.

The times of separation with the above column are: diethylamine, 5 minutes; butylamine, 20 minutes.

Another suitable mixture is propylamine and butylamine.

3. Unsaturated compounds can be separated in the same way as amines using hot water (ca 60 °C) in the titration flask. 0.1 M potassium manganate(VII) (permanganate) is used as the titrant. A suitable mixture for separation is ethanal (acetaldehyde) and propanal. Times of separation are: ethanal, 10 minutes; propanal, 38 minutes.

Gas chromatography as a class experiment

I. W. WILLIAMS

A column is constructed of glass tubing 6–7 mm internal diameter, 80 cm long. This may be in the form of a single U, which may be immersed in a 500 cm³ measuring cylinder; alternatively it may be formed into a two-loop coil [1] which can be immersed in a 600 cm³ beaker. The coil is filled with 'Tide' [2] detergent which has been sieved to remove fine particles. A metal or 'Pyrex' glass jet, about 1 mm diameter, is attached to one end of the column and the other end is connected by means of rubber tubing to the gas mains. The cylinder or beaker is filled with hot water, temperature about 55 °C, ensuring that all the packing in the column lies below the water level.

The gas is turned on and after a minute or so it may be lit at the jet. Using the gas pressure that I have available, the arrangement described gives a small (5 mm) non-luminous flame. If the flame is larger and luminous the coil should be repacked more tightly to reduce the flow.

When a steady, non-luminous flame has been obtained, the gas is turned off and the gas tubing removed. (The only possible danger in the experiment is that pupils may remove the gas lead without turning off the gas tap, for the small flame is not a good indication that full pressure is being applied.) Immediately, a single drop [3] of hexane is introduced from a very fine dropping tube into the column and the gas tubing replaced. The gas is turned on and a stopclock is started simultaneously (alternatively the gas might be turned

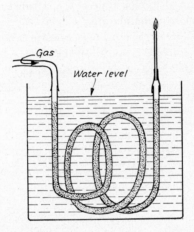

on when the seconds hand of a large laboratory clock passes the minute). The gas at the jet is relit. After about 10 or 15 seconds the flame dies down (due to the air introduced when the gas lead was removed) but is not extinguished and the steady non-luminous flame returns. A short time later the flame becomes luminous, as the hexane is eluted; this persists for a while and then disappears.

The experiment can then be repeated with other hydrocarbons. Some typical results with normal hydrocarbons are:

55–45 °C	pentane	appears 17 sec	dies 28 sec	(B.pt 35.6 °C)
	hexane	30	45	68.8°
	heptane	63	90	98.5°
	octane	180	240	125.6°

The comparison with boiling points is interesting for all pupils. A-level pupils might plot log (appearance time–air time) against the number of C atoms in the hydrocarbon; a straight line is obtained. (Air time = time at which the flame dies down due to the air drawn into the column when introducing the sample.)

The use of the technique for mixtures should now suggest itself. Using three drops of a commercial sample of petrol, a series of peaks of luminosity separated by periods when the flame is non-luminous was obtained:

45 °C	3 drops petrol	luminosity appears 17 sec	dies 20 sec
		35	50
		75	110
		195	270
		510	660

By comparison with the previous results it may be suggested that the first four peaks are due to the normal hydrocarbons and that the fifth may be *n* nonane (extrapolation of the log time plot confirms this). Even better correlation can be obtained by maintaining a fairly constant temperature, such as running the petrol at 50 °C; the temperature factor is not critical, but increasing the temperature does speed up the slower peaks.

Some other interesting hydrocarbons which have been tried on a different column are as follows:

55 °C	*n* hexane appears	43 sec	dies 75 sec	(P.p. 68.8 °C)
	hexyne-1	53	78	71.5°
	benzene	115	195	80.0°
	cyclohexane	100	185	80.9°
	cyclohexene	140	215	83.2°
20 °C	*n* butane	50	60	−0.5°
	n pentane	70	110	35.6°
	n hexane	210	360 ⎫	68.8°
	3-methyl pentane	150	300 ⎬ C_6H_{12}	63.4°
	1,3-dimethyl butane	100	180 ⎭	58.1°

This apparatus will not separate, say, mixtures of the hexane isomers which have overlapping peaks, but the possibilities of refinement by using a longer column are obvious. This would however call for a higher gas pressure— probably from a hydrogen cylinder together with a pressure-regulating device which would be impracticable for class use. As it stands the apparatus gives a clear indication of the principles and power of the method. The minute sample required is striking, but it should be pointed out that this is at one and the same time a strength and a weakness for the technique is difficult to adapt to the bulk separation of mixtures.

NOTES

1. The plane of the coils should be vertical rather than horizontal, for in the latter the packing tends to settle, leaving a channel along the whole length of the packing, whereas in the vertical type settling simply leaves a small dead space at the top of each loop.

2. *Gas Chromatography*, A. I. M. Keulemans, p. 226. Some samples of 'Tide' appear to be rather moist, which results in moisture collecting at the eluting end of the column after a time; this tends to run back into the deter-

gent, dissolving it and blocking the column. The 'Tide' may be dried in an oven at 110 °C overnight, but in most cases this is unnecessary for the slight moisture eluted soon dries up.

3. The major difficulty in the experiment is to introduce a sufficiently small sample into the column. The very smallest drop which can be induced to fall from a fine capillary dropping tube (having a volume of about 0.01 cm³) is sufficient. Slightly more of the higher hydrocarbons and of mixtures are required, say two or three drops. The effect of increasing the sample size is to extend the period over which the flame is luminous; the appearance time is not seriously affected but the persistence time is considerably increased so that in the case of mixtures it becomes impossible to resolve separate peaks.

4. It is found that cheapest grade petrol gives the best results. More expensive grades give less clear-cut peaks due to the presence of additives such as benzene, or a different ratio of hydrocarbons.

5. A further application of the apparatus may also be described:

In distilling coal tar as a demonstration a small fraction was collected at 99 °C, being mainly water with about 3 ml of an organic liquid. Two drops of this liquid were chromatographed on a fairly short column of 'Tide' (25 cm) at 55 °C and periods of luminosity were observed at: 55–125 s, 140–260 s, 390–660 s. In a calibration experiment on the same column benzene gave luminosity from 54–110 s, and toluene from 144–320 s. The third period of luminosity corresponds to the timing that one would expect for xylene and ethyl benzene, the next members of the series. (The presence of toluene and xylene—b.pt 110 °C and 140 °C respectively—in the sample collected at 99 °C is presumably due to steam distillation.) This example provides a a genuine use of the technique to identify quickly (inasmuch as gas chromatography does provide an identification) the components of a rather small specimen.

N.B. These experiments have not yet been carried out using natural gas.

An apparatus for gas chromatography

W. E. ROSSER

More quantitative results may be obtained using the following apparatus:

The column is made out of polythene tubing about 1 m long and 1 cm bore. Polythene tubing is used because it is easy to coil. The column is filled with 'Tide' detergent, which is first heated in an oven at 110 °C to remove any moisture, as this tends to cause a blockage in the column. The coil is then placed vertically in a water bath. We used a thermostat at a controlled temperature of 50 °C.

An early difficulty was the injection of the sample. This was overcome by using a 'Quickfit' stillhead and stopper with B 10 joints (as illustrated). Injection is performed by switching off the gas supply, removing the stopper and placing the sample inside with a fine dropping pipette. The stopper is inserted and the gas supply restarted.

When the sample emerges from the column it passes a coil of platinum wire (15 cm). Another advantage of the polythene tubing is that the wire can be pushed through the tubing when hot. The wire is in a Wheatstone bridge

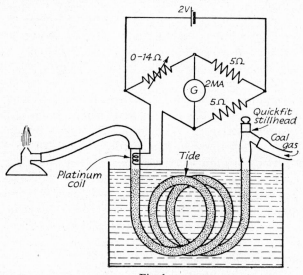

Fig. 1

circuit which has been adjusted for zero deflection when the coal gas alone is flowing. This is done by altering the rheostat indicated. The resistance of the platinum wire depends on the thermal conductivity of the medium that it is in, so that as the sample passes the wire it changes the resistance. The gas in then burnt off at a jet (the base of a Bunsen burner), giving a non-luminous flame about 1 cm high.

Deflection of the galvanometer needle

Deflection	Appears (sec)	Disappears (sec)	Persistence (sec)
1	85	116	31
2	145	185	40
$2\frac{1}{2}$	185	337	152
6	337	367	30
$2\frac{1}{2}$	367	397	30
22	429	599	170
$1\frac{1}{2}$	630	690	60

A sample of commercial petrol was analysed with this apparatus. By measuring the size of the deflection on the galvanometer, the time the deflection first appeared (this identifies which gas is present), and the persistence of the deflection, the results shown on page 76 were obtained.

These figures were plotted as shown in Fig. 2.

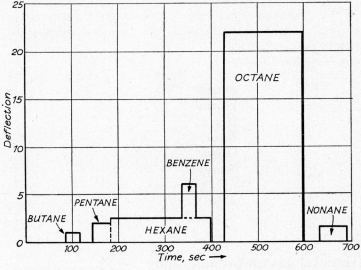

Fig. 2

I. W. Williams (see previous article) showed that it is possible to identify compounds by plotting a graph of log of time of appearance of the deflection against number of carbon atoms in the compound, which gives a straight line. We obtained the graph (Fig. 3) using propanone (acetone), pentane, and octane in the apparatus. Propanone took 63 sec to show a deflection.

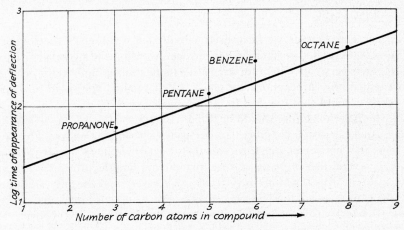

Fig. 3

We could not at first identify the area marked in Fig. 2 as benzene. It was suggested by my colleague Mr D. I. Williams that it might be benzene, so benzene was run through the apparatus and the appearance, time and deflection corresponded to the unknown area.

This does not appear at the time one would expect to fit in with the second graph.

Finally the percentage of the various components of petrol was estimated by measuring the area under the graph for each, and working it out as a percentage of the total area.

The quantities found were:

Butane	0.7%
Pentane	1%
Hexane	13%
Benzene	2%
Octane	83%
Nonane	2%

Experimental gas chromatography

D. E. P. HUGHES

A gas chromatography apparatus has five main parts: the supply of carrier gas together with the necessary pressure control; an injection system; a column and heating system; a detector and a means of showing the output from the detector. Each part will be considered in turn.

THE CARRIER GAS

In all our experiments we have used a hydrogen cylinder as the source of carrier gas; oxygen is unsuitable, but nitrogen would make a good alternative. Hydrogen is very convenient for use with a flame ionization detector provided that adequate fire precautions are taken. We have had one slight mishap in six years' use and that was due to carelessness.

The hydrogen cylinder should be fitted with a fine adjustment valve as well as the usual expansion chamber. In our early experiments we took great care to keep the hydrogen pressure constant by providing a steady blow-off through several feet of water and by introducing lengths of capillary tubing in the system. We now find these to be unnecessary and the only precaution we take is to measure the hydrogen pressure on a mercury manometer. It is quite easy to keep the pressure constant to within a few millimetres when working at 40 centimetres above atmospheric pressure.

THE INJECTION SYSTEM

We use cheap rubber serum caps obtainable through the usual laboratory suppliers. The simplest way to mount them is in a T-junction (Fig. 1). The

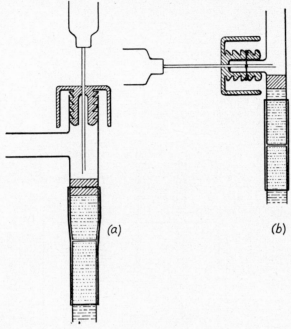

Fig. 1

position of the cap is mainly one of convenience. Method A makes replacement of the cap easier, but sometimes the hypodermic needle is hard to push in: with method B it is essential to wire on the cap. (All rubber-to-glass joins on the high pressure side of the column should be secured in this way.) A serum cap may last up to 100 injections as long as a fine No. 26 needle is used.

Most experiments can be done by using gases or by injecting the vapour of a volatile liquid contained in a bottle or test-tube. For these experiments a cheap plastic syringe works admirably. If liquid samples have to be used, it is possible to make do with an ordinary glass syringe and inject just one drop of liquid. The volume depends on the needle size but is usually about 5 μl. For accurate work there seems to be no alternative to the expensive commercial microlitre instrument.

THE COLUMN AND HEATING SYSTEM

Practical column materials are glass and copper tubing. Plastic and rubber must be avoided as they strongly absorb the injected substance; if they are used for connections it is essential for the glass–glass or glass–copper surfaces to be in close contact.

Very little serious work can be done unless the column can be heated and kept at constant temperature (within a few degrees). Our first column was made from a 1m length of glass tubing bent into a U-tube and heated electrically with nichrome wire wound round the outside; the arrangement is

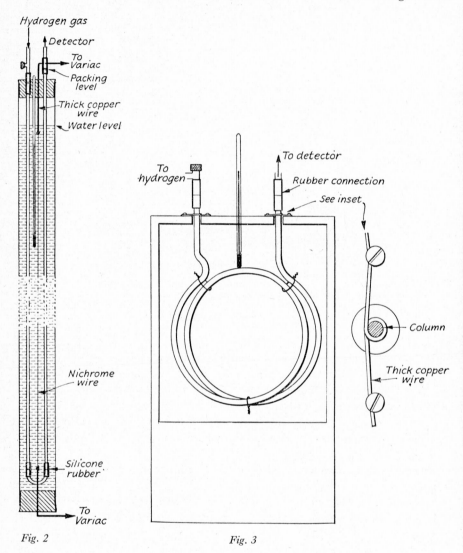

Fig. 2

Fig. 3

not very satisfactory as it is impossible to measure the column temperature accurately. At present we have a 3 m glass U-shaped column for temperatures up to 90 °C (Fig. 2), and a 2 m copper spiral suitable for experiments at 150 °C (Fig. 3). The heating element in Fig. 2 is a 1 m length of 26-gauge nichrome wire heated from an 8-A Variac. By adjusting the voltage it is possible to keep the temperature to within a couple of degrees up to the boiling point of water. We are using silicone rubber connections to the small U-tube at the bottom of the column, but there is no reason why rubber should not make a satisfactory alternative as long as it is replaced fairly frequently. The tubing is 4 mm internal diameter.

The spiral copper tubing is heated in a thermostatically controlled oven; the temperature fluctuates about five degrees between switching on and off. The $\frac{1}{2}$ cm copper tube is fixed by thick copper wire to the screws used for securing the gas vents at the top of the oven. It is essential that at least $2\frac{1}{2}$ cm of the copper tube protrudes between the oven and the rubber connections, or the latter will soon disintegrate. It is possible to make a permanent connection with 'Araldite', but the glass is then very easily broken. It would be much better to have the join made with a proper brass union.

A popular packing is 'Tide'; we find that the separation is much improved by using dinonyl phthalate on Celite. At temperatures above 100 °C, a nonvolatile stationary phase such as silicone oil is essential. We dissolve 15 cm³ of the oil in 100 cm³ of ether and make a slurry with 35 g of 30–80 mesh Celite. The ether is evaporated on a radiator in front of an open window and the powder heated for an hour 30–40° above the intended working temperature.

There is no problem about packing the U-column. The copper spiral is filled before bending. The tube is annealed in a blow-lamp and packed as soon as it is cool. There is then little difficulty in making a 30 cm coil, using a biscuit tin as a former.

THE DETECTOR

Our first experiments were with a thermal conductivity detector, based on a design given by ICI [3]. We found that if the 200 watt filaments had been used they were very fragile; we obtained some new ones from one of the bulb companies and mounted two of them in glass tubes (Fig. 4), submerged in

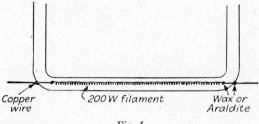

Copper   200 W filament   Wax or
wire   Araldite

Fig. 4

water to keep the temperature constant. A reasonable output was obtained with 0.1 cm³ of sample, but this quantity of liquid flooded the column and gave poor resolution.

Much better results are obtained with a flame ionization detector [4].

Originally we burnt the hydrogen at a platinum-tipped glass jet but never succeeded in eliminating sodium ions from the flame; we had no trouble when we tried out the suggestion of a member of the third-year sixth to use a hypodermic needle. The latest model (Fig. 5) can be constructed for a few pence in a very short time. It is hoped to make a flame ionization detector

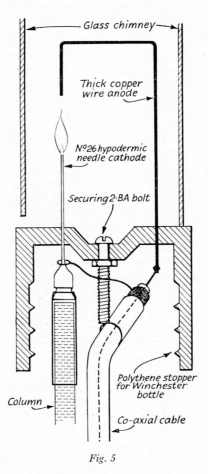

Fig. 5

with an air filter which should remove the dust particles which give 'kicks' to the output recorder. The effect can be minimized by smoothing the output with a large electrolytic condenser (see below).

DISPLAY FROM THE DETECTOR

The Wheatstone Bridge circuit for use with the catharometer has been described elsewhere [3, 5] and will not be considered further.

A flame ionization detector needs a polarizing voltage and must be followed by a very high impedance amplifier as the current flow is only of the

order of 10^{-9} A. With a sensitive amplifier it is possible to use a very small
polarizing voltage (6 V) and feed the output into a 50 MΩ resistor (Fig. 6a):
if the output is very small it could be increased by changing to 300 volts and
150 MΩ respectively. Some people obtain better results by using an alterna-

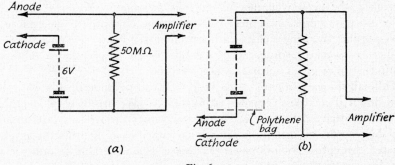

(a)   (b)

Fig. 6

tive arrangement (Fig. 6b); it is then essential to keep the battery contained
in a polythene bag.

A very simple amplifier can be made from a single battery triode with much
reduced anode (and possibly heater) voltage (Fig. 7). The input impedance
of the valve may then be as high as 10^{11} Ω. A large condenser (100 μF)
placed across the meter helps to smooth out the background 'kicks' due to
dust particles ionizing in the flame.

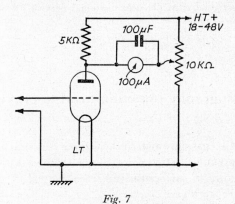

Fig. 7

We have also used the school pH meter as amplifier. The output from the
50 MΩ resistor can be plugged directly into the electrode socket. The instrument is very sensitive but also responds to every flicker.

A most satisfactory amplifier is that described by the Mullard Educational
Service [6] which can be built for a few pounds and is certainly cheaper than
any commercial electrometer amplifier now available. We provide smoothing

by earthing each side of the balanced output via a 1000 μF condenser; an AVOmeter makes a good output instrument as it has a dead-beat movement.

The taking of many meter readings can be most exacting. Two people can soon learn to take readings every five seconds, one giving out the time and the other taking the readings; with practice they can improve this to second intervals if the person taking the readings does not write down the actual values but puts a cross on a piece of graph paper opposite the position of the meter needle. By using a clock with a pronounced tick it is even possible for one person to take readings every second. There is also the possibility of having the times given out over a tape recorder.

While all the experiments to be described can be done in this way, there is no doubt that a pen recorder is invaluable. We are using a Heathbuilt EUW-20A attached to the electrometer amplifier; best results are obtained with the recorder on maximum sensitivity with the output from the amplifier adjusted through a potentiometer. There is a considerable amount of zero drift during the first half hour of operation; after that, full 25 cm deflection can be obtained from a 1 μl sample, with a background noise level of 0.50 cm.

REFERENCES

1. Jaques, D., *S.S.R.*, 1965, 161, **47, 93**.
2. Williams, I. W., *S.S.R.*, 1964, 156, **45, 402**.
3. *The Construction of a Simple Gas Liquid Chromatography Apparatus* (Imperial Chemical Industries Limited).
4. Hughes, D. E. P., *J. Chem. Ed.*, 1965, 8, **42, 450**.
5. Frost, R. J., *Ed. in Chem.*, 1966, 5, **3, 224**.
6. *A low voltage electrometer* (Mullard Educational Service, 1964), Educational Electronic Experiments, No. 8.

A gas chromatography injection head to prevent leakage

T. P. BEAUMONT AND G. KING

Having constructed a gas chromatography apparatus similar to that described by Hughes (see previous article), we found that we were losing hydrogen through the puncture in the serum cap when we used the needles supplied with the 'microlitre' syringe.

We decided to use borosilicate glass, and since the working of this was beyond our competence, the head was constructed for us by Messrs Northern Scientific (York), 2 The Crescent, York. Their model was based on a 2–3 mm bore Springham Interkey stopcock (with retaining ring), replacement keys being available.

Stopcock T has a bore of 2–3 mm (enlarged if necessary with a length of copper wire coated with grinding paste). It is important that the tap should be fitted with some retaining device to prevent displacement of the barrel. One arm (A) is opened out to accept a 9 mm Suba-Seal cap. The distance H should be less than the length of the syringe needle (4 cm will probably be found satisfactory). The other arm (C) has a sidearm (B) joined to it as close to T as possible.

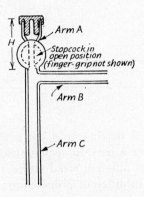

Whilst the column is in use, T is in the closed position and carrier gas passes from the cylinder via B and C to the column. To inject the sample, T is opened and the syringe needle passed through the bore. After injection, the needle is withdrawn and T closed again.

If a 'microlitre' syringe is not available, the problem of introducing a sufficiently small sample may be overcome by using a teat pipette drawn out to a fine capillary, which is a sliding fit in a small hole made in the Suba-Seal cap (or other suitable rubber closure). The capillary may, with care, be inserted through the bore and the sample expelled on to the column. This operation is facilitated by lightly lubricating the edges of the hole with silicone grease.

We measure the flow rate of the carrier gas using a 'Meterate' tube B with glass float. This is one of a series of flowmeters supplied by Messrs Jencons Scientific Ltd, of Hemel Hempstead. The seven tubes in the series (each of which can be fitted with either a glass or stainless steel float) cover a wide range of flow rates for gases and liquids.

Experiments with the gas chromatography apparatus

D. E. P. HUGHES

RETENTION TIMES

Any preliminary experiments with the apparatus must emphasize the basic principle of gas chromatography: that a single substance is retained by the column for a characteristic time which is independent of the sample size or the presence of other substances.

Each substance gives a peak which ideally should be symmetrical (Fig. 1), though often there is 'tailing'. In every case the retention time is taken from

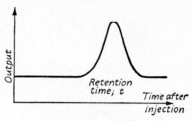

Fig. 1

the time of injection to the maximum of the peak and can be found quite easily, even though a meter is used to display the output.

It is instructive to inject different sized samples of a pure liquid and also a mixture of two liquids in different proportions: suitable pairs are normal and 2-methylbutane, methyl and ethyl ethanoates (acetates) and benzene and cyclohexane. Under favourable conditions the retention times should be reproducible within 5 per cent, and 0.1 per cent of one substance be detected in the presence of the other.

During these experiments particular attention must be given to temperature and pressure control. The latter is difficult to keep constant from day to day and it is better to measure the flow rate of the carrier gas with a simple soap bubble device [1, 2], and convert times into retention volumes

$$V = t.F$$

where F is the flow rate in cm^3/sec (assuming the time to be in seconds). We have found that with careful pressure control, it is usually unnecessary to convert to retention volumes in this way, though for really accurate work it is essential.

(i) Retention times for a homologous series

In gas–liquid chromatography the retention time is dependent on the distribution coefficient of the substance between the carrier gas and the non-volatile liquid on the column. It is likely that these distribution coefficients vary in a regular way for a homologous series and that there is smooth relationship between retention time and number of carbon atoms in the series. Fig. 2 shows this for the paraffin hydrocarbons.

Samples of methane and ethane can be made by standard laboratory methods, propane and butane from gas lighter fuel (three main peaks, the middle being 2-methylpropane), while pentane is available in bottle. Other homologous series which could be studied include the ethanoate esters, aldehydes or ketones.

The graph in Fig. 2 does not pass through the origin. The intercept on the time axis shows the time taken for the carrier gas alone to pass through the

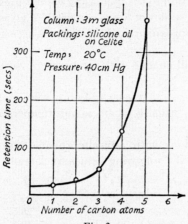

Column: 3m glass
Packings: silicone oil
on Celite
Temp: 20°C
Pressure: 40cm Hg

Fig. 2

column: this is known as the 'hold-up' time t_0. The actual time taken for the substance to pass through the column is given by

$$t_r = t - t_0,$$

where t_r is called the adjusted retention time. A plot of log t_r against number of carbon atoms gives a straight line (Fig. 3).

The hold-up time can often be found directly as the injection of the sample interrupts the smooth flow of carrier gas, and gives a small deflection on the

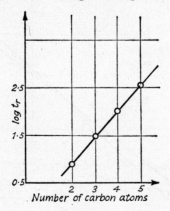

Fig. 3

meter. Alternatively the hold-up time may be assumed to be equal to the retention time of a very low boiling point, non-polar gas such as methane. We used this method for finding t_0 for Fig. 3 and hence methane does not appear on the graph.

(ii) Variation of retention time with temperature

If the temperature of the column is raised, the adjusted retention volume V_r will decrease in proportion to the increasing vapour pressure of the substance. For a given substance, a plot of log t_r against $1/T$ gives a reasonably straight line with a slope of approximately $L_v/(2.3R)$. There are two main errors in this simple treatment:

(*i*) the flow rates will change with temperature, so that V_r is not directly proportional to t_r;

(*ii*) the gas expands, so that each retention volume should be adjusted to the same temperature, usually taken as the temperature of the measuring instrument.

In an experiment done with 2-bromopropane over a temperature range from 20 °C to 120 °C, the following values for latent heat were obtained:

Method	L_v kJ mol^{-1}
t_r	23.1
V_r	26.5
V_r corrected to 20 °C	30.7

The last value agrees closely with the value expected from the Trouton Rule. The experiment is a good exercise in the careful presentation of results.

(iii) Variation of retention time with boiling point

A very useful calibration curve can be drawn up by plotting log t_r against boiling point for a range of substances with similar polarity. Similar types of

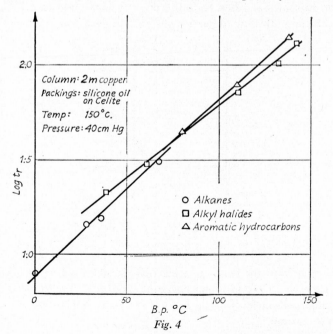

Fig. 4

substances often give parallel straight lines, but the slope changes for a very different class of compounds (Fig. 4).

An unknown substance may be tentatively identified by estimating its boiling point and looking up boiling point tables. It cannot be emphasized too strongly that this method can be totally unreliable unless a check is made on the retention time with a pure sample of the suspected compound. Even if the retention times agree there can still be ambiguity because several different substances can have similar retention times.

In practice unknown peaks are identified from standard tables of retention volumes; for school use the boiling point method is much more simple but must be used with great care.

(iv) Quantitative analysis

It is difficult to do much quantitative work without an expensive microlitre syringe. One possibility is to make a 1 per cent solution of the substance in a volatile solvent such as ether and use an ordinary syringe; this method is only suitable for liquids with a relatively high boiling point. If the substance is very volatile, gas injection can be used. In order to obtain reproducible results it is essential to use a standard technique; a simple one is to have a little of the liquid in a corked test-tube at room temperature, invert once and then insert the syringe and needle into the vapour above the liquid.

For small samples the height of the peak is approximately proportional to the sample size (Fig. 5). Equal quantities of different substances do not of course give the same deflection. Isomers give the same area under the peak but not the same deflection as the higher boiling isomer gives a fatter maximum. The sensitivity of a catharometer depends on the thermal conductivity and hence the molecular weight of the substance. The output from a flame

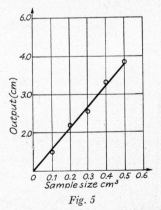

Fig. 5

ionization detector is more variable; it is thought to be related to the number of carbon ions formed in the flame. This means that the output for equimolar samples should be proportional to the number of carbon atoms in the molecules.

Other experiments

It is also possible to try the effect of different pressures across the column. The retention time will become smaller as the pressure is increased, but the volumes should remain the same for not too large pressure changes.

It would also be instructive to try columns packed with materials of differing polarity, for example silicone oil, dinonyl phthalate and polyethylene glycol-400. Substances of similar boiling points (for example hexane and 1-bromopropane) should then be eluted in the following order: polar on non-polar, non-polar on non-polar, polar on polar. With a suitably polar column it is easy to separate benzene and cyclohexane, even though their boiling points differ by only one degree.

ANALYSIS OF MIXTURES

Analysis of familiar substances

Once a gas chromatograph is established as a familiar piece of laboratory equipment, there are frequent requests to analyse common liquids such as lighter fuel, cleaning solvent, model aircraft fuel, spirits or petrol.

The town gas supply provides an interesting trace (Fig. 6); hydrogen and carbon monoxide are absent as they do not ionize in the flame ionization detector.

Analysis of laboratory chemicals

It is instructive to show pupils that 'pure' chemicals from the bottles are often quite complex mixtures. It is not surprising that ordinary samples of

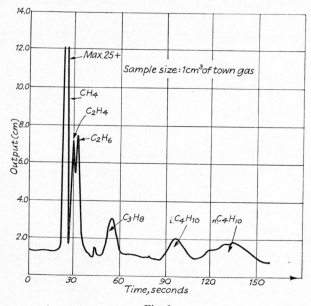

Fig. 6

pentane invariably contain detectable amounts of 2-methylbutane, but the presence of some impurities in other substances is less predictable. For example 1,1-dichloroethane contains about 1 per cent of at least four different substances which are probably various propyl and butyl chlorides. One can then ask: why were these substances formed during manufacture, and why were they not removed during subsequent purification? The difficulty of purification is emphasized if the mixture is redistilled with a laboratory fractionating column; there is usually not much difference in the proportions of the impurities present.

APPLICATIONS

Undoubtedly the real value of gas chromatography in schools is in applying the technique to help solve problems from all branches of chemistry. The following list of suggestions is not meant to be exhaustive; it is given in order to indicate the possibilities that are available for project work.

The purity of laboratory preparations

Fig 7 shows a trace obtained from a laboratory-prepared sample of bromo-ethane before and after purification. During the course of a double practical period six pairs of traces were obtained, all showing similar features. The first peak is easily identified as ethene and the last double hump is ethoxy-ethane (diethylether) and bromoethane. The retention times in each case agreed closely with the values for pure samples. The second peak was unexpected and discussion with the class brought forward the suggestions of methoxymethane, methoxyethane and bromomethane. The peak corresponds to a substance with a boiling point of about 5 °C and so eliminates the possibility of methoxymethane. The final identification was done by repeating the run at the lower temperature of 20 °C and using as standards samples of bromomethane and methoxyethane prepared in the laboratory. The retention time of the unknown peak was the same as that of the bromomethane, and with a large sample a very small peak (less than 0.1 per cent) corresponding to methoxyethane was also detected. This experiment has been described in some detail as it does show how a routine laboratory preparation can form the basis of a most fruitful investigation.

Organic reactions

During the last decade gas chromatography has been extensively used to elucidate organic mechanisms. A convenient reaction to study is the Wurtz synthesis. The reaction is known to go smoothly with higher alkyl halides although the lower ones are often given as examples in textbooks. We have not yet detected any butane from bromo- or iodoethane, either in solution or in the vapour phase. I-bromopropane gives a fair yield of hexane though we have never approached the quoted value of 50% [3]. It might be possible to try the Wurtz reaction under different conditions and see if there is any evidence to support the idea that it proceeds by a free radical mechanism in the vapour phase [4].

Many other organic systems would be worthy of study; two possibilities are the action of nitric(III) acid (nitrous acid) on amines and the variation in o, m and p ratios in aromatic substitution. Another suggestion is the action of heat on heavy metal salts of organic acids [5].

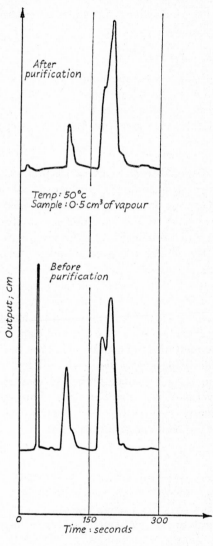

Fig. 7

Physical chemistry problems

Rapid gas analysis makes it possible to carry out experiments on vapour pressures. After initial calibration with a liquid sample, vapour pressure/ temperature curves can be found. Mixtures of liquids can be used to illustrate Raoult's Law: for example a 1 cm³ sample of the vapour above an equimolar mixture of 1- and 2-bromopropanes gives two peaks of the same height as those from 0.5 cm³ samples of the separate liquids. An extension of this idea is to determine the efficiency of a fractionating column [6]. These experiments will only give reasonable results if great care is taken to clean out the syringe when injecting vapours of different composition; the problem is particularly acute with plastic syringes and with these a clean syringe should be used for each injection and the dirty ones put in a warm place to bake out.

Gas chromatography could also be applied to rates of reaction and equilibrium constants particularly in the gas phase. The rates of decomposition of 2-chloro-2-methylpropane, isomerization of cyclopropane and the photochemical halogenation of paraffins are possible suggestions. A simple gas phase equilibria might be that between butane and 2-methyl propane in the presence of aluminium chloride as a catalyst; the classic nitrogen dioxide–dinitrogen tetroxide system could not be used with a flame ionization detector as it does not respond to inorganic compounds.

REFERENCES

1. Kaiser, R., *Gas Phase Chromatography I* (Butterworth, 1963), p. 78.
2. Frost, R. J., *Ed. in Chem.*, 1966, 5, **3**, 226.
3. *Nuffield Chemistry: A course of options* (Longman/Penguin Books, 1967), p. 136.
4. Sykes, P., *A guidebook to mechanism in organic chemistry* (Longman, 1961), pp. 202–3.
5. Barnard, J. A. and Chayen, R., *Modern methods of chemical analysis* (McGraw-Hill, 1965), pp. 184–6.
6. Ault, A., *J. Chem. Ed.*, 1964, 8, **41**, 432.

Index